Proceedings of the First National Expert and Stakeholder Workshop on Water Infrastructure Sustainability and Adaptation to Climate Change

U.S. Environmental Protection Agency
Office of Research and Development
Office of Water

April 2009
EPA-600-R-09-010

Table of Contents

Acknowledgements iii

Disclaimer iii

Acronyms iv

1.   Introduction 1

1.1   Overview of the Workshop 1

1.2   EPA Office of Research and Development: Commitment to Climate Change Research and Adaptation 2

1.3   The EPA National Water Program: Sustainable Water Infrastructure and Adaptation to Climate Change 3

2.   Challenges and Opportunities in Adapting to Climate Change 4

2.1   Adaptation Challenges to the Nation and the Science Community 4

2.2   Perspectives from Utilities 5

2.3   When R&D Meets the Real World: The Challenges and Opportunities of Integrating Water Resource Management for a Changing Climate 7

3.   Applying Climate Science to Water Infrastructure Planning 9

3.1   Information Needed for Infrastructure Adaptation Planning 9

3.2   Where the Research Meets the Road: Climate Science, Uncertainties, and Knowledge Gaps 11

3.3   Holistic ORD Research to Ensure Water and Energy Efficiency through Drinking Water System Sustainability 12

3.4   Accommodating Design Uncertainties: Past Practices and Future Needs 13

4.   Research and Development for Water Infrastructure Adaptation 15

4.1   EPA's Global Climate Change Science Program and Water Infrastructure Adaptation Research 15

4.2   AWWARF Research Strategy for Climate Change Adaptation 17

4.3   WERF's Climate Change Research Programs 18

4.4   Incorporating Climatic Uncertainties into Water Planning 19

5.   Climate Change Impacts on Hydrology and Water Resource Management 22

5.1   Projecting Hydroclimatic Changes – Downscaling 22

5.2   Projecting Hydroclimatic Changes – Local Applications of Downscaling 27

5.3   Evaluating Hydoclimatic Change for Water Infrastructure Adaptation – Part I 32

5.4   Evaluating Hydoclimatic Change for Water Infrastructure Adaptation – Part II 39

6.   Adaptive Management and Engineering: Information and Tools 45

6.1   National Infrastructure Condition Assessment and Adaptability 45

6.2   Progressive Adaptation: Planning and Engineering for Sustainability 52

6.3   Adaptation Practices and Tools – Part I 58

6.4   Adaptation Practices and Tools – Part II 64

7.   Moving Forward in Adaptation 71

7.1   Concluding Remarks 71

7.1   Suggested Ideas and Recommendations for Moving Forward in Adaptation 72

Appendix A   List of Workshop Participants **98**

Appendix B   Workshop Agenda **107**

Appendix C   Biographies of Workshop Speakers and Moderators...............................**112**

Acknowledgements

The U.S. Environmental Protection Agency (EPA) gratefully acknowledges the contributions of the following individuals for successfully conducting its First National Expert and Stakeholder Workshop on Water Infrastructure Sustainability and Adaptation to Climate Change:

EPA Workshop Coordinating Committee

> Y. Jeffrey Yang, Office of Research and Development, NRMRL
> Karen Metchis, Office of Wastewater Management, WPD
> Elizabeth Corr, Office of Ground Water and Drinking Water, DWPD
> Jill Neal, Office of Research and Development, NRMRL
> Robert Cantilli, Office of Science and Technology, HECD

EPA Workshop Support

> Michael Shapiro, Office of Water
> Sally Gutierrez, Office of Research and Development, NRMRL
> Steve Heare, Office of Ground Water and Drinking Water, DWPD
> Linda Boornazian, Office of Wastewater Management, WPD
> Suzanne Rudzinski, Office of Science and Technology
> Jeff Peterson, Office of Water
> Jim Goodrich, Office of Research and Development, NRMRL
> Tom Speth, Office of Research and Development, NRMRL

Support from Abt Associates Inc. and Stratus Consulting, Inc. under EPA Contract EP-C-07-023

Audio and Transcription Services from The Track Group

Disclaimer

Proceedings of the First National Expert and Stakeholder Workshop on
Water Infrastructure Sustainability and Adaptation to Climate Change

iii

Acronyms

AMWA	Association of Metropolitan Water Agencies
ASCE	American Society of Civil Engineers
ASR	Aquifer Storage and Recovery
AWWA	American Water Works Association
BMP	Best Management Practice
BYWO	Build Your Way Out
CAT	Climate Assessment Tool
CCAR	California Climate Action Registry
CLIMB	Climate's Long-Term Impacts on Metro Boston
CSO	Combined Sewer Overflow
CSS	Combined Sewer Systems
CWA	Clean Water Act
DBP	Disinfection By-Product
DOE	U.S. Department of Energy
EBMUD	East Bay Municipal Utility District
EPA	U.S. Environmental Protection Agency
FACA	Federal Advisory Committee Act
FEMA	Federal Emergency Management Agency
GCM	General Circulation Model
GCM	Climate Change Model
GDP	Gross Domestic Product
GEOSS	Global Earth Observation System of Systems
GHG	Greenhouse Gas
GIS	Geographic Information Systems
HSPF	Hydrologic Simulation Program-Fortran
ICLUS	Integrated Climate and Land Use Scenarios
ICR	Information Collection Request
IDF	Intensity-Duration-Frequency
IPCC	Intergovernmental Panel on Climate Change
IWRM	Integrated Water Resources Management
KBDI	Keetch-Byram Drought Index
LID	Low Impact Development
MBR	Membrane Bioreactor Reactor
MPDI	Modified Perpendicular Drought Index
MWAI	Metropolitan Water Availability Index
MWRA	Massachusetts Water Resources Authority
NACWA	National Association of Clean Water Agencies
NAWC	National Association of Water Companies
NCAR	Natural Resources Defense Council
NRMRL	National Risk Management Research Laboratory
NWP	National Water Program
O&M	Operations and Maintenance
ORD	Office of Research and Development
ORSANCO	Ohio River Valley Water Sanitation Commission

Proceedings of the First National Expert and Stakeholder Workshop on
Water Infrastructure Sustainability and Adaptation to Climate Change

iv

OW	Office of Water
PCB	Polychlorinated Biphenyl
PDSI	Palmer Drought Severity Index
POTW	Publicly Owned Treatment Work
PUC	Public Utilities Commission
PV	Photovoltaics
R&D	Research and Development
RCM	Regional Climate Models
RIO	Ride It Out
SAM	Strategic Asset Management
SDWA	Safe Drinking Water Act
SRES	Special Report on Emissions Scenarios
TDS	Total Dissolved Solids
TMDL	Total Maximum Daily Load
TOC	Total Organic Carbon
TTHM	Trihalomethane
USDA	U.S. Department of Agriculture
UV	Ultraviolet
WEF	Water Environment Foundation
WEPP	Water Erosion Prediction Project
WEPPCAT	WEPP Climate Assessment Tool
WGA	Western Governors' Association
WRAP	Water Resources Adaptation Program
WRF	Water Research Foundation
WSMP	Water Supply Management Program
WSS	Water Supply and Sanitation
WTP	Water Treatment Plant
WUCA	Water Utility Climate Alliance
WWTP	Wastewater Treatment Plant

Proceedings of the First National Expert and Stakeholder Workshop on
Water Infrastructure Sustainability and Adaptation to Climate Change

ii

1.   Introduction

The U.S. Environmental Protection Agency (EPA) held its First National Expert and Stakeholder Workshop on Water Infrastructure Sustainability and Adaptation to Climate Change on January 6-7, 2009, in Arlington, Virginia. Sponsored by the EPA Office of Water and Office of Research and Development, the workshop was attended by more than 130 invited experts and stakeholders from the federal, research, utility, engineering, academic, and NGO sectors. A list of attendees is provided as Appendix A.

Speaker panel from plenary session moderated by Dr. Pei-Yei Whung, EPA office of the Science Advisor

The workshop included several plenary sessions, as well as two concurrent tracks:

- Climate Change Impacts on Hydrology and Water Resource Management

- Adaptive Management and Engineering: Information and Tools

The agenda is provided in Appendix B. These proceedings include summaries of each of the presentations, as well as the discussion sessions. Where available, hyperlinks are provided to each of the presentations on the EPA Web site. For each session, hyperlinks to the transcript of the presenter's remarks are provided (with the exception of concluding remarks in Chapter 7). Appendix C includes biographies for the speakers and moderators.

1.1   Overview of the Workshop

Jim Hanlon, Director, EPA Office of Wastewater Management

Cynthia Dougherty, Director, EPA Office of Groundwater and Drinking Water

Jim Hanlon and Cynthia Dougherty opened the workshop by welcoming the participants. Climate change will have a large effect on water utilities and EPA is exploring ways to help them adapt and manage water infrastructure. During this workshop, we will learn what tools and methods are needed to maintain and improve water infrastructure. The infrastructure built today will be in place for decades to come, and infrastructure planning decisions are being made each day. EPA and utilities need to find better ways to maintain the current water infrastructure and prepare for changes. EPA's Sustainable Infrastructure Initiative fits well into this discussion.

Important changes will affect water resource management in coming decades. A National Research Council (NRC) study states that 42 percent of urban land areas will be redeveloped by 2030. The United States is facing population growth of 20 percent by 2030 and 40 percent by 2050. EPA and utilities need to look at green infrastructure and water reuse, for example, especially with the upcoming economic stimulus package in which major investment decisions will be made. This workshop is an opportunity to reach out to experts to build the agenda to deal with climate change in the context of adaptation.

Proceedings of the First National Expert and Stakeholder Workshop on Water Infrastructure Sustainability and Adaptation to Climate Change

1

It is very important to gain a better understanding of what EPA can do to help utilities make decisions and where EPA can make investments in science and research. It is also important for EPA to understand what the utilities are already doing, and how to engage in research that is complementary, not redundant.

The focus of this workshop is on precipitation-related impacts. Although sea-level rise is an important concern for coastal utilities, this topic will be only touched upon here and will be left for a more complete discussion at a future workshop. Also, this workshop is focused on adaptation. While the interaction between water utilities and energy is critical (water utilities use 3 to 4 percent of total U.S. energy), this workshop will not address mitigation efforts.

It is important to note that this workshop is not being held under the Federal Advisory Committee Act (FACA), and therefore EPA is not looking for a consensus among participants. Further, there are a number of federal partners involved in this workshop, and EPA is looking forward to collaborating with them.

Click here to read the transcript of Mr. Hanlon's and Ms. Dougherty's remarks.

1.2 EPA Office of Research and Development: Commitment to Climate Change Research and Adaptation
Sally Gutierrez, Director, ORD National Risk Management Research Laboratory

The EPA National Risk Management Research Laboratory (NRMRL) is conducting research on the country's aging water and wastewater infrastructure. The goals are to protect and improve public health, save energy, and improve the capacity for water infrastructure systems to incorporate sustainability into their planning. NRMRL has begun to focus on climate change adaptation in the context of aging infrastructure. This focus has led to the development of the Water Resources Adaptation Program (WRAP). WRAP is developing assessment and adaptation decision support tools, as well as water resources and infrastructure adaptation and multi-scale assessments (national and regional adaptation strategies). NRMRL is working with the Office of Water to coordinate the development and dissemination of data and tools.

The current toolboxes include drinking water treatment and water availability forecasting. NRMRL's objectives in this area include discussing science needs, developing adaptation tools, and helping develop interim decision-making processes. This includes fostering collaborative research and accelerated transfer of research and development (R&D). NRMRL hopes to elicit feedback and recommendations on EPA's climate change R&D activities and national water program.

Click here to view Sally Gutierrez's presentation.

Click here to read the transcript of Ms. Gutierrez's remarks.

Proceedings of the First National Expert and Stakeholder Workshop on Water Infrastructure Sustainability and Adaptation to Climate Change

2

1.3 The EPA National Water Program: Sustainable Water Infrastructure and Adaptation to Climate Change

Benjamin Grumbles, Assistant Administrator, EPA Office of Water

The National Water Program (NWP) recognizes that water is at the heart of the debate on climate change. Two years ago, the NWP established a task force on climate change headed by Mike Shapiro and Jeff Peterson. This task force developed a strategy to provide a connection between science and resource management. The focus is not only on mitigation, but on adaptation strategies and policies.

Consistent with the strategy, this workshop focuses on adaptation and the steps that need to be taken to educate stakeholders and build solid partnerships. This workshop was not designed to simply restate the current science on climate change, but to provide EPA and utilities a connection between the science and resource management, and to focus not just on mitigation but also on adaptation. This workshop should help improve the communication between EPA and water managers on what the water managers can do about climate change under drinking water and clean water laws. Leaders in all areas need to work together and engage in conversation focusing on adaptation, education, and management. For this reason, EPA has a strong partnership with states, other utilities, and everyone involved in this effort.

EPA recognizes that much of the discussion on adaptation relates to infrastructure. To continue on the sustainability theme, it is important to emphasize resiliency. Gray infrastructure has to be able to last 30 to 70 years. This workshop will also look at programs, standards, and water management techniques. One of the keys to success is to ensure that scientists and managers are working together.

Federal agencies are working together on adaptation. In August 2008, Under Secretaries from various federal departments and agencies began discussions regarding key areas related to water and climate change. EPA has worked with the U.S. Department of Agriculture, the Department of the Interior, the Army Corps of Engineers, and the Commerce Department, and sent a directive to agency colleagues to being coordinating and sharing information. This workshop will help inform those discussions.

When members of the American public hear the term "climate change," they are concerned about it but not necessarily thinking about water. If they are, they are usually thinking about sea-level rise or melting glaciers. They don't yet understand the implications that climate change has for water resources and the need to adapt to the impending changes. This workshop should help inform regulators and infrastructure managers to make better decisions with respect to climate change and water infrastructure.

Society faces "insurmountable opportunities." The Chinese word for crisis is composed of two symbols that represent danger and opportunity. Ben Grumbles closed his remarks by stating, "Ready. Set. Adapt."

Proceedings of the First National Expert and Stakeholder Workshop on
Water Infrastructure Sustainability and Adaptation to Climate Change

3

2. Challenges and Opportunities in Adapting to Climate Change

In this session, speakers discussed the challenges and opportunities faced by the water management and science communities when considering how to adapt to climate change. The concept that "stationarity" is dead' brings into question many of the standard tools and practices that have been adopted. This is a particular challenge as the water management community must make investment decisions now for expensive infrastructure that will have a long life span. The research community is grappling with how to develop tools and approaches that can aid water managers in taking future climate change and a changing hydrology into consideration in decision making.

Click here to read the transcript of the remarks of the moderator (Dr. Pai-Yei Whung, EPA Office of the Science Advisor).

2.1 Adaptation Challenges to the Nation and the Science Community
Peter Gleick, Pacific Institute

Peter Gleick began his keynote speech by stating that society needs to "manage unavoidable impacts, and avoid unmanageable impacts." The impacts of climate change are unavoidable and occurring right now. The debate regarding climate change is entering a new phase, where scientists and policy makers must develop appropriate and effective responses that will address both mitigation and adaptation. Recommendations to water managers on climate change have been available for two decades, but progress toward implementing these recommendations has been slow. In many cases, climate change poses serious challenges to managed water systems. Relying on current engineering practices may lead to incorrect and potentially dangerous decisions.

Dr. Gleick made several recommendations. First, the 2000 National Assessment Report on Water should be updated. Climate change needs to be integrated into all management and planning activities at the federal, state, and local levels. Eight years ago, the Secretary of the Interior issued an order to integrate climate change into all decisions, but the order was ignored. There is also a need to improve assessments of the energy/water connections. Three to four percent of total U.S. energy use goes to the water industry, and this percentage is much higher when home energy use for water is included (e.g., heating). Therefore, water utilities will have to think about mitigation efforts. In California, utilities and researchers are starting to assess the energy and greenhouse gas footprint of water utilities.

There is a need to expand the concept of infrastructure to include nonstructural measures, such as low-flow toilets, and precision irrigation or computer-controlled distribution canals for farmers. It is important to think about infrastructure as more than dams, aqueducts, and wastewater treatment plants. There is plenty of money available to upgrade traditional infrastructure, but if we expand the definition of water infrastructure to include nonstructural measures on the water side as well, investments could be more efficient. Also, if these nonstructural measures were included in the concept of infrastructure, many jobs could be generated through the development of efficient appliances and the installation of toilets and washing machines that use water efficiently.

When reviewing the science on impacts of climate change, there are projections of hotter temperatures, uncertain changes in precipitation, dramatic changes in snowpack and melt, sea-level rise with its impacts on aquifers and delta ecosystems, and extreme events (e.g., flooding and drought). The models show that the western United States is expected to lose a significant amount

Proceedings of the First National Expert and Stakeholder Workshop on Water Infrastructure Sustainability and Adaptation to Climate Change

4

of snowpack. This has enormous implications for water management in California. Snowpack changes have not yet led to managers changing their reservoir use. Instead, there is talk of building more reservoirs to deal with snowpack loss.

The concept of infrastructure needs to be expanded to include non-strucutural measures, such as low-flow toilets or precision irrigation systems or computer-controlled distribution canals. In addition to upgrading traditional infrastructure, emphasizing these other measures would be an efficient means to conserve the water supply and a wise investment. The need for adaptation is not new for California; the state has already experienced changes in the water cycle.

Dr. Gleick closed his talk by noting that the water community has for years recognized that adaptation to climate change is needed. In 1997 the American Water Works Association's (AWWA's) public advisory forum recommended that water agencies consider climate change, but they are still not doing so. Policies should integrate and coordinate mitigation and adaptation, and managers should review the advantages and disadvantages of existing water management policies. Useful climate change models are available to deal with variability, but they may not be enough. It is important to explore ways to incorporate adaptation into planning, and develop and test adaptation strategies. Examples of incorporating adaptation include testing and modifying operating rules, redefining infrastructure, improving water-use efficiency, and better evaluating the energy/water/greenhouse gas (GHG) links. Integrating climate change into water planning will help managers understand existing risks and exposure. Managers and planners can no longer assume that hydrologic conditions will look the same in the future as they have in the past. Even the best models are of little value if managers do not integrate climate change into water management decisions.

Click here to view Dr. Gleick's presentation.

Click here to read the transcript of Dr. Gleick's remarks.

2.2   Perspectives from Utilities
David Behar, San Francisco Public Utilities Commission

David Behar explained that the Water Utility Climate Alliance (WUCA) began at a San Francisco conference in the winter of 2007. WUCA is a consortium of water providers, and its mission is to improve research, develop strategies for adaptation, and reduce emissions from water utilities. WUCA has buy-in from the member utilities' general managers, and each member utility has senior staff who focus on adaptation, holding conference calls on adaptation issues every several weeks. WUCA has developed an awareness of climate change impacts on water and identifies system vulnerabilities to these impacts. It integrates climate change risk assessment into strategic and capital planning and informs ratepayers of the science and potential costs. Key discussions include how best to integrate risk assessments and address what is known and not known regarding climate science.

WUCA also addresses how to make decisions on adaptation, and how best to influence policy and climate change science research and investment, particularly at the federal level. The Alliance is thus seeking to hold conversations about these issues with climatologists and agency leaders. However, WUCA finds that these are often difficult conversations to have, because in many cases, the engineering and utility management focus is different from the day-to-day focus of the agency leadership, which in turn is different from the day-to-day focus of the climate science community.

Putting the above-mentioned discussions in the context of capital investment is vital. Mr. Behar described two capital investment projects that have taken place in the context of climate change: (1) San Diego County Water Authority: A $4.3 billion capital improvement program that includes enhanced storage capacity, desalination, and a new water treatment plant; and (2) San Francisco Public Utilities Commission: A $4.3 billion project that includes a pipeline under San Francisco Bay, dam replacement, and recycled water and groundwater facilities.

In Reports to Congress (2005, 2008), EPA has estimated that investment needs in drinking water total up to $277 billion, and pollution control investment needs are an estimated $203 billion, for a total of $480 billion. These investments need to be made in the context of climate change. For example, San Francisco is rebuilding a dam that is a 150-year investment. These investments will last well into the future, when there will be significant climate change effects. When capital investment decisions are being made, it is important to understand how the investments could influence climate change policy.

The other key challenge is the significant uncertainty in predicting the implications of climate change in order to make efficient capital investments. In particular, there is significant uncertainty about regional precipitation changes in the future. Varying projections in the Northwest make planning very difficult. Seattle is looking at a 6 to 21 percent decrease in water supply, and San Francisco is facing much earlier snowmelt than in the past. These precipitation changes are uncertain, although newer models tend to show a decrease in total precipitation.

WUCA calls for a comprehensive federal effort on predictive climate change tools. They want to see improvement in the quality and accessibility of regional modeling (also known as downscaling). WUCA has several efforts underway, such as a climate modeling white paper that will review the state of the science on climate modeling and analyze where the most productive investments can be made. The paper also addresses when to expect more useful climate change projections. WUCA is also working on a decision support white paper, being developed by Malcolm Pirnie, which will survey the decision support tools currently in use.

WUCA calls for "actionable science," which includes data, analyses, and forecasts that are sufficiently predictive, accepted, and understandable to support decision making, including capital investment decision making. WUCA is focusing on climate-driven strategies as opposed to what are known as "no regrets strategies."

WUCA is taking direction from other organizations such as the Western Governors' Association (WGA), which in 2008 stated that water managers should clearly communicate their needs to the science research community. The WGA said that water managers should focus on getting ahead of research, decisions, and policy.

Mr. Behar concluded by providing the next steps for the water utility community. These steps include recognizing the need for "actionable science," and considering the needs of water utilities before research agendas are set, not after. There is a need for a coordinated and strategically directed federal science effort and an increase in resources and computer modeling capability.

Click here to view Mr. Behar's presentation.

Click here to read the transcript of Mr. Behar's remarks.

Proceedings of the First National Expert and Stakeholder Workshop on
Water Infrastructure Sustainability and Adaptation to Climate Change

6

2.3 When R&D Meets the Real World: The Challenges and Opportunities of Integrating Water Resource Management for a Changing Climate

Dr. James Goodrich, EPA ORD National Risk Management Resource Laboratory

James Goodrich said that the EPA Office of Research and Development (ORD) is looking at applied and practical research and wants to focus the research down to the local level. We are seeing changes in precipitation and intensity, leading to more runoff and erosion. Not everyone will face the same challenges; what Philadelphia and Boston do to adapt will be far different from what cities such as San Diego and Dallas do. Many cities across the United States have the same number of people in them, but have followed very different growth paths. Some have boomed in recent years, while others have depopulated. How those cities adapt to climate change will be impacted by their growth patterns.

There are several formidable challenges, such as whether society knows enough to adapt. How does society know how much adaptation is needed? What are the uncertainties in predictions, and what is the best way to deal with them? What methods and techniques are available, and what can science and engineering do? Is the climate variation natural or anthropogenic? This workshop will attempt to look at the methods and techniques that EPA, utilities, and other organizations are currently using. It will be important to figure out the adaptation challenges that society is facing.

Dr. Goodrich presented several research questions and topics:

- Downscaling and how to use it locally by providing tools and methodologies for local managers,

- Engineering information and tools,

- Planning and engineering of water infrastructure for sustainability,

- Predicting impacts on hydrology and water quality in watershed scales, and

- Design criteria that consider energy, land use, and population changes.

In conducting this research, it important to bear in mind that the water industry is conservative, as it makes decisions that affect communities for many decades. Technological and institutional changes are needed to orient the systems toward sustainable water services. The centralized approach of big pipes in and out has worked well in the past, but will not work in the future. There could be a point where water quality returns to pre-EPA conditions if water systems stay on their current path. Industry needs to look at new materials, real-time monitoring, and new energy sources, and needs to evaluate its carbon footprint, but must do these things at a reasonable cost. There will likely be a shift toward multiple treatment systems or a decentralized approach focusing on reuse and recycling.

The EPA Water Resources Adaptation Program (WRAP) methodology is the integration of climate change, hydrological response, and land use infrastructure. This is combined with EPA's Sustainable Infrastructure Initiative, which is particularly looking at vulnerability assessments. The combined efforts of the two programs will ultimately help the industry develop new, sustainable water use infrastructure. Dr. Goodrich concluded by reiterating that the many federal agencies, universities, financial institutions, and water research foundations need to have a coordinated R&D approach.

Click here to view Dr. Goodrich's presentation.

Click here to read the transcript of Dr. Goodrich's remarks.

Summary of Discussion Session

A representative from the Water Research Foundation (WRF) stated that climate change adaptation strategies have been around for two decades. WRF has conducted an exercise to review 400 WRF projects since 2002. Of these 400, 130 were related to climate change; however, only eight had climate change in their titles.

A water manager mentioned that he often sees great possibilities for R&D, and then R&D fails to quickly move to implementation. The big questions are, "Who is impacted by action or inaction? Who benefits and who pays?" The current financial model does not match up with the available solutions. Science might decide the most cost-effective outcomes, but right now, the finances do not add up. Great projects such as water infrastructure are likely to be put aside because of other investment needs. It is important to find the right financial model.

A water resources researcher said that economics is not being addressed in either of the two workshop tracks, but could be discussed in Track B (Adaptive Management and Engineering). He thinks there is currently enough money, but managers and policy makers still do not integrate climate change into investment decisions. This is not just an economic issue. King County, Washington, recently looked into retrofitting homes rather than invest in a bigger wastewater treatment plant. County staff found that it was cheaper to retrofit the homes, but they built the new plant anyway due to risk factors.

A local government water manager stated that costs are a crucial factor. As important as it is to receive federal funding, such funding is not likely to materialize. Although it is important to fund a cap-and-trade system for adaptation, it may not happen. More money will have to be found to meet additional needs. California Governor Arnold Schwarzenegger and Les Snow wanted to build additional dams in anticipation of climate change, but were told they could not have the money. The local government manager has not found any infrastructure projects that anticipate climate change except for backflow devices in San Francisco. The argument must be made to ratepayers about the future effects of climate change. For instance, if utilities are going to raise rates 10 percent, they need to make a good argument for the increase.

A water resources researcher asked how to minimize the effects of raising rates. A local water manager answered that a good argument is needed for the rate increases. In San Francisco, the rate increase had to be decided by public referendum. A water resources researcher stated that money is not the biggest problem. In many cases, the available funds have to be spent anyway. The biggest problem is getting climate science into the policy arena. The information that he thinks is "actionable" is not what utilities think is "actionable." A participant from the engineering community said that the most important factor is that the consumer base is not valuing water properly. A local water manager added that customer education and perception regarding desalination and water reuse is a key factor in getting approval for capital improvements. In the past, projects have been held up due to protests by small groups.

Proceedings of the First National Expert and Stakeholder Workshop on
Water Infrastructure Sustainability and Adaptation to Climate Change

8

3.   Applying Climate Science to Water Infrastructure Planning

Building on the themes identified in Session I (Section 2), this session delved into specific details such as decision-making approaches and the use of climate models, data, and information to aid in incorporating climate change into water infrastructure planning. Perspectives were given from the water management, consulting, and research communities.

Click here to read the transcript of the remarks of the moderator (Jim Taft, Association of State Drinking Water Administrators).

3.1   Information Needed for Infrastructure Adaptation Planning
Stephen Estes-Smargiassi, Massachusetts Water Resources Authority

Stephen Estes-Smargiassi said that the mission of the Massachusetts Water Resources Authority (MWRA) is to provide an adequate and reliable supply of high-quality drinking water. In addition, its goal is environmentally responsible collection, treatment, and disposal of wastewater. However, funds for this are always an issue because customers place higher priority on the utility delivering water all the time and under any circumstances.

MWRA has several strategies for dealing with climate change. These strategies include (1) improving regional climate change projections, (2) enhancing the understanding of potential impacts, (3) determining and implementing appropriate adaptations, (4) inventorying and managing GHG emissions, and (5) improving communications and tracking mechanisms.

Mr. Estes-Smargiassi stated that the take-away message from his presentation is, "It's just engineering." Water utilities already think about risk and consequence and do not need to be convinced to act responsibly. While not everyone is convinced that climate change is happening, all want utilities to act responsibly. In addition to physical adaptation, standard procedures will also need to change as a result of climate change. It is important to consider how regulations and guidelines the industry currently operates under will need to change, and also how the underlying legal structure will need to change.

Mr. Estes-Smargiassi discussed scenario planning versus probability-based planning. Stationarity was a great concept in the past, but while assumptions and coefficients are changing, the underlying problems and physical processes are not. A flood is still a flood, pipes still must carry water, and rain still falls and evaporates. There must be a plan for inundation from larger storms as a result of climate change. Because buildings and equipment have a limited lifespan, and it is important to ensure that each facility has planned and invested appropriately for the long term. This investment needs to fit into the current maintenance and upgrade cycles, and the results must be reviewed on a regular basis. There is also a need to change design curves.

Planners must follow the flood all the way "upstream" to determine needed adaptation. Data that planners need include new design flood elevations (the Federal Emergency Management Agency (FEMA) should issue new flood maps) and the frequency and distribution of storm intensity. Planners must be appropriately conservative, since three meters of sea-level rise will require flood-proofing all of Boston, rather than just MWRA facilities—an assumption that would involve prohibitively high investment costs. However, planners must figure out where to draw the line for the medium-term investment.

Proceedings of the First National Expert and Stakeholder Workshop on
Water Infrastructure Sustainability and Adaptation to Climate Change

9

Mr. Estes-Smargiassi discussed a recent project that was designed with climate change in mind. The design planners of the Deer Island Wastewater Treatment Plant (WWTP) were challenged to think about higher rises in sea level. The plant was raised two feet for the 50-year forecast. An analysis was done and some costs rose, but the planners took a guerrilla approach. It is important to note that no one in management knew that the planners did this. This decision was made almost 20 years ago, and it is not clear in light of current sea-level rise projections if the plant was raised enough.

Further data needed include storm frequency, precipitation runoff, and more geographically discrete data. It will be vital to put the new information into guidance and design manuals and to get engineers to start working with the updated data.

Two other key issues to focus on include a change in regulatory guidance and a change in customer expectations. There is also a need to acknowledge the hierarchy of risks, for example (homes, yards, streets, shellfish resources, recreation, and finally water use bodies). Planners need to think about base demands as the population shifts, seasonal demands change, and access to emergency supplies from surrounding jurisdictions shift or change. Multiple resources must integrate into multiple regions (known as regionalization). By focusing on risks and consequences, managers already plan and invest based on events that are unlikely to occur, such as a 1,000-year event. However, singling out climate change has made it harder for decision makers to make investment decisions. Climate change should be rolled into the set of risks that are already being evaluated. There is no need to convince every decision maker regarding climate change effects, but it is important to treat climate change like all of the other risks that water managers face.

Laws and regulations are not stationary; just as the physical environment adapts to climate change, laws and regulations must adapt as well. Each year, there may be a need to change agreed-upon solutions under a policy on combined sewer overflows (CSOs) (e.g., the acceptable number of CSOs). Utilities shield themselves from liability by using generally accepted standards; for example, the utility is deemed negligent if there is a failure under typical storms, but not under an extreme event.

The National Center for Atmospheric Research (NCAR) advocates using the theory of decision making under uncertainty to enact laws and regulations that adapt to climate change. This process will rely more on applying probability theory and more intensive modeling. Therefore, there is a strong need for probability-based analysis. However, wildcards include increased variability, extremes, and large population movements. Extreme weather events are likely to be more intense, which can lead to exactly the type of conditions that could affect water supplies and drainage systems. There should be a focus on research needs to better understand these risks and this variability.

Mr. Estes-Smargiassi concluded that most demographic projections assume stationarity, but questions if this is a valid assumption. There could be major discontinuities, such as retirees returning to the Snow Belt, or the long-term livability of coastal cities. An additional question for planners pertains to whether or not the discount rate should be set at zero. It is difficult to plan for the long term with a rate greater than zero.

Click here to view Mr. Estes-Smargiassi's presentation.

Click here to read the transcript of Mr. Estes-Smargiassi's remarks.

Proceedings of the First National Expert and Stakeholder Workshop on
Water Infrastructure Sustainability and Adaptation to Climate Change

10

3.2 Where the Research Meets the Road: Climate Science, Uncertainties, and Knowledge Gaps

Dr. Dennis Lettenmaier, University of Washington

Dennis Lettenmaier stated that there is a need to integrate climate science much better into the engineering process. The Seattle Public Utilities Commission (PUC) updates its plan every five years, but has not considered climate change at all, even though it has funded many studies on this subject. The reason for this is that while public policy analysts knew about climate change, engineers did not understand what to do about climate change. Engineers have focused only on stationarity.

There are several issues facing utilities in planning for climate change. Design and management are based almost entirely on analysis of historic observation, risk, and reliability (e.g., based on a 100-year event). Methods are often standardized; for example, the Water Resources Council Bulletin 17b is used for estimating flood risk. However, utilities do not have standardized methods when they do not assume the climate is stationary. Dr. Lettenmaier declared that it is therefore important to think about probability analysis.

Dr. Lettenmaier discussed the methods of probability analysis that are available to planners and engineers. General Circulation Models (GCMs) provide the best information. In the past, GCMs have been used only to create scenarios. There is a need to address GCM differences as a representation of uncertainty and incorporate them into the planning process. Each GCM run is a representation of what the climate system might do. The difference between individual models can be considerable. Since the climate system is chaotic, it is better to perform multiple runs of models and multiple models to create ensembles, rather than rely on just one or a few models.

It is important to extract information from GCMs for planning, while considering the following:

- Bias is a key issue,

- All GCMs are not created equal,

- It is not clear how to weight GCMs, and

- Multiple ensemble models should be used.

Dr. Lettenmaier discussed the methods available for downscaling. It is important to note that downscaling methods reproduce GCM uncertainties and will not resolve disagreement among GCMs. Statistical downscaling is easier to apply than dynamic methods and can give a better representation of uncertainty because downscaling methods can be simulated for hundreds to thousands of years. Regional climate models (RCMs) have more realistic topography than GCMs. A major limitation is that they are typically run for a ten-year simulation, a period that is very short for risk analysis.

Climate change needs to be incorporated into water planning in a routine manner. There is also a need for a standard for the archiving of data sets to allow easy access to historical information. Planners must recognize that climate projections will be updated, which may mean that they move away from the critical planning period and incorporate climate change into the engineering process.

Dr. Lettenmaier concluded by mentioning that applications research needs to include:

- A better understanding of the elements of uncertainty in GCM ensembles,

- Better archiving of data sets,

Proceedings of the First National Expert and Stakeholder Workshop on
Water Infrastructure Sustainability and Adaptation to Climate Change

11

- A stronger push for applied R&D on ways to incorporate hydrologic research into water engineering, and

- A need for a more systematic approach to regional climate simulations and downscaling.

Click here to view Dr. Lettenmaier's presentation.

Click here to read the transcript of Dr. Lettenmaier's remarks.

3.3 Holistic ORD Research to Ensure Water and Energy Efficiency through Drinking Water System Sustainability
Dr. Audrey Levine, EPA Office of Research and Development

Audrey Levine described the evolution of water infrastructure, which was originally designed strictly as a supply system for fire fighting. Over time, it evolved into a source of potable and non-potable water. Wastewater and stormwater collection then evolved to divert drainage and control health risks. Today, treatment systems are evolving toward developing better treatment, discharge, and reuse techniques. Preventing oxygen depletion and improving water quality are two of the main goals of modern systems. There are currently 160,000 drinking water systems in United States serving more than 300 million people. The United States relies on these systems, which are of varying ages and conditions, to provide water for consumers that expect it to be safe for drinking and other applications.

There are several current and emerging concerns, including the sustainability of water systems, which includes availability, infrastructure, public health, and competing demands. Another concern is the energy/water interdependencies that emerge when moving and treating water and managing infrastructure. Energy policy impacts water in a number of ways, and there are growing concerns with biofuels and geological sequestration and their respective water demands. As society moves more toward water resource replenishment, augmentation, and restoration to meet in-stream and downstream uses, more integrated water systems will be needed.

Dr. Levine described several climate change research drivers, including water availability and water quality (including evaporation, microbiology, pathogen diversity, and survival). Other research drivers include energy and economic impacts on treatment reliability, including availability of chemicals, transportation issues, restoration of the hydrologic cycle, and sustainable use for ecosystems. Finally, water is not currently recovered and reused consistently. It is vital to preserve high-quality water and make this water easy to transport.

It is important to be able to balance water availability and water use to restore and maintain the hydrologic cycle. Priority must also be placed on the subject of water and ecosystem sustainability. Dr. Levine discussed several further research issues to focus on, including public health protection. More research is needed on ways to optimize collection treatment systems for water quality, water recovery, and energy efficiency. Green infrastructure, low-impact development, integration of centralized and decentralized systems, and water economics are also key research areas. Dr. Levine concluded by stating that ORD's drinking water research program is focusing on several long-term goals, including characterizing and managing risks and organizing the program around the hydrologic cycle.

Click here to read the transcript of Dr. Levine's remarks.

*Proceedings of the First National Expert and Stakeholder Workshop on
Water Infrastructure Sustainability and Adaptation to Climate Change*

12

3.4   Accommodating Design Uncertainties: Past Practices and Future Needs
Doug Owen, Malcolm Pirnie Inc.

Doug Owen began by noting that the engineering community faces fundamental questions: "where is the water, when are we going to get it, how much are we going to get, what form is it going to be in, and what's the quality?" Mr. Owen noted that changing temperatures will have additional impacts on regional planning, behaviors, and ecosystem health. Engineers have developed processes and approaches for developing reliable designs that enable infrastructure to function under a range of conditions. With climate change, engineers now face the additional question of how they can accommodate climate change uncertainties into sustainable water infrastructure.

Engineered systems are conduits between inputs and outputs. They balance short-term requirements and long-term needs within an economic framework. Fluid flow and hydraulics are based on physical laws and are well understood by the engineering community, but the major uncertainty lies in the inputs and outputs. The conventional approach to engineering infrastructure includes selecting the appropriate planning horizons, evaluating the alternatives, selecting the preferred alternative, and designing the project using standard engineering principles. Engineers use appropriate and cost-effective conservatism to account for uncertainty. The uncertain inputs and outputs include population growth and demand. The data for inputs and outputs include urban development, households and employment, income, water prices, and conservation. It is vital to have those data before developing the engineering solution. Historical records have traditionally been used to assess variations in natural phenomena and usually include 75 to 100 years of records. There is now a movement toward using more predictive models. Climate change will affect concentrations of pollutants as a function of both rainfall and runoff and may require a different solution set.

Mr. Owen said that design conservatism has two forms, safety factors and redundancy. Advantages of safety factors include reliability and possible excess capacity as operating efficiencies are better understood. Disadvantages include the fact that engineers may strand assets and systems may not operate efficiently, particularly if the design is too large. Engineers are always balancing between designing as efficiently as possible and as flexibly as possible. The advantages of redundancy include protecting against unit failure and decreased need for unit maintenance. Disadvantages include the possibility that one unit may not be operating at any given time and the added cost to the system. Costs drive smaller redundant modules so that there is not as much excess unused capacity.

Engineers must decide how to manage safety factors as a function of the consequence of failure. Costs increase with more safety factors, while consequences are often difficult to predict and cost. The consequences are often pre-determined by regulation or by community need. These include regulations and needs regarding drinking water, wastewater effluent, and combined sewer overflows. Infrastructure cannot be quickly changed to incorporate changes to regulations, needs, and/or consequences. It can often take 15 years or longer from project evaluation to operation. Historically, planning horizons have provided some degree of conservatism. But as changes occur, this conservatism may strand excess capacity, which may lead to designing systems in smaller increments. Approaches also can differ for above- and below-ground facilities. Engineers do not want to dig below ground because digging can represent up to 75 percent of infrastructure costs. Engineers can also put in bigger pipes at a marginal cost; however, if the system is oversized, this can lead to water quality deterioration and septic and pipe degradation over time.

Mr. Owen stated that analyzing the "knee in the curve" regarding, for example, the building of combined sewer overflows (CSOs) can maximize incremental benefits. However, climate change will alter where the knee occurs; thus, managers must balance infrastructure and operational investment and try to stay at the lower end of the cost curve. Energy prices and climate change will affect design and process selection through a shift to renewable sources, lifecycle assessment, and energy audits or operational modifications. Mr. Owen said that in order to make the best use of existing infrastructure and determine where to invest, engineers need a regional understanding of impacts and knowledge of how precipitation patterns and temperature will change. He concluded by saying, "some people see the glass half empty, and some people see the glass half full, while an engineer sees the glass twice as big as it needs to be."

Click here to view Mr. Owen's presentation.

Click here to read the transcript of Mr. Owen's remarks.

Proceedings of the First National Expert and Stakeholder Workshop on
Water Infrastructure Sustainability and Adaptation to Climate Change

14

4. Research and Development for Water Infrastructure Adaptation

The water management and research communities are developing information and tools that can aid water managers in adapting to climate change. Representatives from EPA, WRF, the Water Environment Research Foundation (WERF), and a water utility discussed research and tool development strategies that are underway and what is on the research agenda for future development. The presentations helped identify research gaps and can be used to develop agendas for research and tool building for water infrastructure adaptation.

Click here to read the transcript of the remarks of the moderator (Carol Collier, Delaware River Basin Commission).

4.1 EPA's Global Climate Change Science Program and Water Infrastructure Adaptation Research
Dr. Joel Scheraga, EPA Office of Research and Development

Joel Scheraga began by stating that information and tools are already available to incorporate climate change into decision making. The EPA/ORD Global Change Research Program has a well-defined mission to provide timely and useful scientific information to support decision making. The primary focus is to assess the potential consequences of global change, particularly climate variability and change, in the United States.

The focus areas of the ORD program are water quality/aquatic ecosystems, air quality, and human health. The program focuses on adaptation research to reduce risks and to take advantage of opportunities presented by global change. The program is stakeholder-oriented, involving collaboration with decision makers in particular locations to support decision making by trying to understand what research is needed and when it is needed. The program is integrated across all of EPA's laboratories and centers.

The program is currently trying to understand how climate change will affect EPA's ability to fulfill its statutory, regulatory, and programmatic requirements (e.g., the Clean Air Act, Clean Water Act, and Safe Drinking Water Act). The ORD program is undertaking a major assessment of climate change on water quality as part of EPA's well-defined role within the multi-agency U.S. Climate Change Science Program (CCSP). This role includes assessing consequences for decision makers, evaluating adaptation options, and developing decision support tools. In its 2001 research strategy, EPA established a goal to assess the potential impacts of global change on water quality and aquatic ecosystems in the United States. Major assessments under this goal included:

- Evaluation of the consequences of global change for water quality related to pollutants and microbial pathogens (2005-2006).

- Development of the BASINS decision support tool for incorporating climate variability and change into water management decisions (2007).

- Evaluation of the consequences of global change for stream and river biological indicators (2007).

- Evaluation of the effects of climate change on aquatic invasive species and implications for management and research (2007).

Proceedings of the First National Expert and Stakeholder Workshop on
Water Infrastructure Sustainability and Adaptation to Climate Change

15

- Ongoing development of information and tools on global change impacts and adaptation options in key watersheds.

Dr. Scheraga described several success stories, including a study of combined sewer systems (CSSs). Three to four years ago, EPA Region 5 and Great Lakes mayors approached ORD about looking at CSSs. There are 770 CSSs that serve about 40 million people. The study looked at whether climate change matters in the redesign of systems. It found that climate change may result in a failure to meet the established standards. There could be an average of 237 events per year above the control policy's objectives across 182 communities. Key conclusions of the study included:

- Climate change will affect future performance of many CSSs in the Great Lakes region.

- Calculations of system size should not be based on current hydrology and historic precipitation data.

- A policy decision must be made about additional investments to build in a margin of safety.

- The risks posed by climate change to CSSs are manageable, and this also provides an opportunity to link with smart growth policies.

The Global Change Research Program is also developing user-friendly decision support tools such as enhancing BASINS with the Climate Assessment Tool (CAT) (available at http://www.epa.gov/waterscience/BASINS/). CAT will help determine how water resources could be affected by a range of potential changes in climate, and the effectiveness of management practices for increasing the resilience of water resources to changes in climate.

The program is now undertaking a major water quality assessment with the Office of Water (OW). The Global Program will work with OW to conduct a study of the sensitivity to climate change of goals articulated in the Clean Water Act and the Safe Drinking Water Act, and the opportunities available within the provisions of these acts to address the anticipated impacts. The planned water quality assessment will be incorporated into the new OW climate change strategy. Future planned activities include:

- An assessment of OW needs and priorities relating to water quality and global change.

- Broad-based, national scale assessment of water quality endpoints vulnerable to global change.

- Detailed watershed-based, stakeholder-driven studies focused on local issues and specific management solutions for addressing global change.

- Detailed studies of the potential impacts and opportunities for adapting water infrastructure and the built environment.

- Development of broadly applicable decision support tools to increase the capacity of OW clients to assess and manage the impacts of global change on water and watershed systems.

Dr. Scheraga concluded by mentioning several reports that are planned:

- Report on the potential of sustainable/green infrastructure to increase resilience to global climate change (2011).

- Report investigating the adaptation techniques (e.g., water reuse) and advanced water conservation approaches to increase infrastructure resilience to global climate change (2012).

Proceedings of the First National Expert and Stakeholder Workshop on
Water Infrastructure Sustainability and Adaptation to Climate Change

16

- Synthesis report on the adaptive potential of water resources development and water infrastructure engineering and management for responding to global climate change (2013).

Click here to view Dr. Scheraga's presentation.

Click here to read the transcript of Dr. Scheraga's presentation.

4.2 AWWARF Research Strategy for Climate Change Adaptation
David Rager, Greater Cincinnati Water Works

David Rager introduced the climate change strategic initiative of the American Water Works Association Research Foundation (AWWARF, which is now known as the Water Research Foundation or WRF). The Water Research Foundation is one of the world's largest nonprofit organizations doing research in critical drinking water research. WRF has more than 900 subscribers, and is largely a utility subscriber-based organization. Most of the research is driven by the needs identified by the utilities. The foundation has been in existence since 1960 and has completed more than 1,000 research projects for more than $450 million in research. The foundation identified four objectives under its climate change strategic initiative.

WRF's goals include: (1) improve industry awareness of climate change, (2) provide a set of tools to assess vulnerability and develop adaptation strategies, (3) develop tools for assessing water utilities' carbon footprint, and (4) communicate information to internal and external stakeholders.

WRF works globally with other organizations and has held research workshops in Edinburgh, London, and Denver. These workshops identified research priorities, such as water resources, water quality and treatment, infrastructure, energy and environment, and communications and management. There are also five key focus areas that the Global Water Research Coalition, of which WRF is a member, developed: water resources, water quality treatment, infrastructure, energy and environment, and communications and management.

Mr. Rager mentioned several projects the foundation is currently working on, including: (1) climate change impacts on the regulatory landscape and evaluating opportunities for regulatory change, (2) analysis of changes in water use in regional climate change scenarios where anticipated water demands and use patterns under a range of climate change scenarios are being examined, and (3) a project focusing on ground water quality impacts resulting from geologic carbon sequestration.

Mr. Rager identified two issues for water utilities regarding climate change. The first issue is energy management, in which utilities are looking to lower their energy use. The second issue is the impacts of climate change on water utility viability, particularly in regards to financing. He noted that utilities have concerns about water conservation because of the need for future rate increases.

Planned research for WRF includes climate change impacts on the regulatory landscape; vulnerability assessment and risk management tools; analysis of changes in water use and regional include climate change scenarios; change of mindsets to promote design of sustainable infrastructure that includes climate change; and carbon sequestration and its effects on utilities. Research not related to climate change (though still affected by climate change) includes energy management and desalination. Utilities are seeking answers to these issues, and WRF is trying to lay the groundwork for cooperation.

Proceedings of the First National Expert and Stakeholder Workshop on
Water Infrastructure Sustainability and Adaptation to Climate Change

17

Click here to view Mr. Rager's presentation.

Click here to read the transcript of Mr. Rager's remarks.

4.3   WERF's Climate Change Research Programs
Claudio Ternieden, Water Environment Research Foundation

Claudio Ternieden said the first priority for the Water Environment Research Foundation (WERF) with regards to climate change research is to learn what others are doing. No one organization can do it all alone; cooperation is key to moving forward on climate change issues. Once there is an understanding of what is being done, it is important to share information on lessons learned. Another priority is to develop tools needed to make management decisions.

WERF's desired long-term outcomes of climate change adaptation research include opportunities to identify and reduce GHG emissions and nitrogen removal. It is important to minimize impacts to operations due to changing hydrologic and climate conditions. Another long-term goal is to make decisions on capital improvements in the face of uncertainty. This is where developing decision-making tools is extremely important, focusing on asset management. Finally, WERF wants to communicate management approaches and their costs to customers.

WERF's long-term operations optimization goals include reducing the environmental footprint of waste water treatment plants (WWTPs) and improving solids management practices (i.e., for biosolids). WERF hopes to facilitate breakthroughs of innovative and emerging technologies, improve resource recovery, minimize energy use, and shift from energy consumption to a renewable energy production paradigm. WERF wants to take a holistic approach to operations optimization to adapt to climate change.

In 2008, WERF published an assessment of international practices called "State of the Science Report: Energy and Resource Recovery from Sludge." It included technical, capital cost, and operating and management cost information for numerous technologies in various stages of development. The report uses the "triple bottom line" approach to look at social, economic, and environmental considerations. Climate change research issues for WERF include the development of value-added research to provide a solid understanding of the likely impacts of climate change, including impacts on water quality, wastewater services, and costs. Research into developing planning tools and operational management to cost-effectively mitigate and adapt to climate change is another priority.

WERF's "Buyer's Guide to Climate Risk Information" provides guidance on the availability of climate change information for planning. It contains guidance on the use and interpretation of downscaled climate model results, including the latest climate risk information and tools. The guide is a collaborative effort with AWWARF and the UK Water Industry Research (UKWIR). Coordinating with WRF to help fill in research gaps, WERF has recently issued a request for proposals for a research team to develop a white paper to characterize climate change impacts on clean water. WERF expects the paper to be completed by July 2009.

WERF has identified further wastewater climate change research that includes demonstrating sequestration of carbon in biosolids. It is also researching the development of infrastructure planning that can adopt cost-effective responses to climate change. Other areas of research include identifying carbon footprints in new infrastructure, methane emissions from septic systems and force mains, and other fugitive sources. Finally, WERF hopes to identify barriers to better climate change adaptation and mitigation. Mr. Ternieden left the audience with two questions to consider: What are your priorities in wastewater climate change research? What tools will you need to make management decisions in the light of climate change?

Click here to view Mr. Ternieden's presentation.

Click here to read the transcript of Mr. Ternieden's presentation.

4.4   Incorporating Climatic Uncertainties into Water Planning
Marc Waage, Denver Water/Water Utility Climate Alliance (WUCA)

Marc Waage began by stating that traditional water supply planning is based on observed weather and hydrology. It assumes historic variability that history repeats itself and that climate is stationary. This traditional method fails to properly treat the uncertainties from a changing climate. Water utilities performing assessments of their vulnerability to climate change are faced with a large set of future climate projections. There is significant uncertainty about precipitation changes, and water supply and demand is very sensitive to small changes in climate. When utilities try to incorporate climate change projections into planning, they must use a minimum of three climate scenarios.

Traditional planning has relied solely on historical variability. Rather than planning for a small number of outcomes, based on recorded weather and hydrology time series data, utilities have new planning methods available that can use more than 500 climate scenarios with many sources of uncertainty in addition to climate change. Utilities need to accept and plan for this large amount of uncertainty. Mr. Waage introduced the idea of the cone of uncertainty. This concept is used to compare traditional planning with what may be needed in terms of new planning concepts. Uncertainty can grow over time and create a larger cone; therefore, rather than trying to plan optimally for a small set of possible outcomes from a static climate situation, there is a need to plan for robust solutions that work well under a range of possible outcomes over time.

Mr. Waage facetiously presented the seven steps to adaptation, illustrating the importance of uncertainty in new planning concepts: 1. Deny Uncertainty; 2. Debate Uncertainty; 3. Investigate Uncertainty; 4. Attempt to Reduce Uncertainty; 5. Accept Uncertainty; 6. Plan for Uncertainty; and 7. Adapt to Uncertainty.

WUCA decision support objectives include aiding in the transition from stationarity- to uncertainty-based planning methods. WUCA hopes to bridge the gap between projections and the need to make decisions. It aims to identify, understand, and evaluate decision support methods that will allow climate uncertainties to be incorporated into planning. Finally, WUCA hopes to raise awareness of decision support needs and promote research to improve methods. A white paper will be published in April to identify and evaluate different planning methods.

Mr. Waage named four promising planning methods. *Scenario planning* is a systematic way to develop many outcomes in the future. It uses a small number of equally likely scenarios, and in the short run, it identifies commonality among scenarios. Scenarios are more about paradigms (e.g., high water quality, green (environmental) conditions, economic woes). Two case studies on scenario planning were conducted in Tucson, Arizona and Denver, Colorado. Denver Water incorporated scenario planning into its integrated assessment plans.

Proceedings of the First National Expert and Stakeholder Workshop on
Water Infrastructure Sustainability and Adaptation to Climate Change

19

Robust decision making is a computer analysis of many equally likely scenarios. It includes iteration, hedging, and decision points. The method is not optimal, but is more robust than scenario planning. Two case studies on robust decision making were developed for the Inland Empire (California) Municipal Water District and Denver Water.

Decision analysis is an older method that uses decision trees and probabilities and minimizes expected costs. The analysis can apply probabilities from recent climate models.

A *real options* method combines decision analysis and financial theory. It also involves decision tree and financial hedging concepts.

Mr. Waage concluded that traditional planning tools are inadequate to deal with climate change. Utility managers are saying that they are overwhelmed by the amount of uncertainty that climate change represents, but they still want to incorporate climate change into planning. WUCA is conducting research to identify, understand, and modify new methods. It is developing case studies and promoting these methods.

Click here to view Mr. Waage's presentation.

Click here to read the transcript of Mr. Waage's remarks.

Summary of Discussion Session

A water manager raised the issue of carbon expenditures relating to water quality benefits (How many tons of carbon dioxide equivalent are worth water quality benefits?) A water researcher answered that this is a question that will be addressed in the future, but WERF is currently looking at nutrient removal. A climate impacts researcher added that Congress wants to focus on mitigation as well. This question needs to be addressed. His program is looking more at mitigation, and EPA is looking at the ancillary benefits of adaptation investment for mitigation.

A climate impacts researcher asked to what extent research is being coordinated. Another climate impacts researcher answered that within the CCSP is a major effort to look at the appropriate ways to do adaptation research and develop decision support tools. CCSP members are addressing the relative roles of different entities (e.g., federal, state, and private). There are different roles for different organizations, but he does not know the answer because they are currently trying to answer that question. They have test-beds to learn what it takes to adapt effectively to develop decision support tools (e.g., partnership with the Alaska Department of Environmental Conservation to develop their climate change strategy). David Rager added that WERF is a subscriber-based organization and their clients expect work be coordinated. There are laboratories of knowledge in different areas. It is good to consider different ideas, and WERF has received good unsolicited ideas. WERF currently has peer review groups and advisory committees, but there is a need for independent research.

A member of the engineering community offered a third-order problem: uncertainty about uncertainty is uncertain. We should do research on operations because reservoir management can be just as effective as new construction. The conjunctive operations of two reservoirs have provided more reliability than five new reservoirs. Another example given was that a small increase in energy efficiency in Southern Nevada led to more benefits. In New York, a change in operations will do as well in controlling turbidity as in building new infrastructure. A small increase in efficiency will give more benefits than from everything we build in the future.

Proceedings of the First National Expert and Stakeholder Workshop on
Water Infrastructure Sustainability and Adaptation to Climate Change

20

A water manager asked for more information on the Chicago example. A climate impacts researcher answered that EPA made a presentation to the mayors of Great Lake cities. The presentation included a scenario analysis, and it tried to bound plausible climate outcomes, downscaled to the Great Lake region. The analysis produced a robust set of outcomes and all cases had more water quality exceedences. Subsequently, Mayor Daley of Chicago and Chicago's Chief Engineer developed Chicago's smart growth initiative. For example, they expanded combined sewer overflow operations using green infrastructure, began using more permeable materials for paving, and began building culverts better. Chicago recently released a climate change strategy that includes adaptation.

5. Climate Change Impacts on Hydrology and Water Resource Management

This track focused on outputs from modeling efforts by the climate science community that can be used by water resource managers for decision making, primarily identifying climate change information that is now available or may become available within a few years. The sessions focused on methods of developing higher resolution output (particularly precipitation) from climate models through downscaling, as well as quantifying and understanding model output uncertainties, and how model output is being used in decision making. The track also covered how observed climate data can be used to support decision making in conjunction with downscaled model output.

5.1 Projecting Hydroclimatic Changes – Downscaling

One of the difficulties in using climate model projections to support decision making is the low resolution from general circulation models (GCMs). These models project climate change on grid boxes that are typically several hundred miles wide – an areal resolution too coarse for most water resource decision making. Downscaling can produce results with higher resolution. Dynamical downscaling using regional climate models (RCMs) can yield estimates at a scale of tens of miles, while statistical downscaling methods can give point estimates or regional estimates at less than 10 miles. Downscaling typically use probabilistic modeling approaches that incorporate statistical data to enable decision makers to make their decisions based on both the probability of an event as well as the risk or uncertainty associated with that decision. This session reviewed developments in downscaling and how downscaled output could be used in decision making.

Click here to read the transcript of the moderator's remarks.

Downscaling or Decision-Scaling? An Overview of Downscaling
Dr. Casey Brown, University of Massachusetts

Downscaling refers to the process of translating climate projections from coarse resolution GCMs to finer spatial resolution that is considered more useful for assessing local and regional climate change impacts. GCM outputs are for the most part applicable at the "continental scale" with seasonal or annual values. In addition, they have inherent uncertainties. These uncertainties are due to several reasons, including that models typically do not account for the effects of topography and other factors that influence the climate system. Further, correlation with 20th century data does not necessarily imply skill in projecting climate in the 21st century. To account for this uncertainty and the inevitability of model errors, using an "ensembling" approach that generates means and medians from a suite of models often provides higher levels of projection confidence.

Downscaling methods are typically categorized as statistical or dynamical. The statistical method is generally simpler and more efficient, using statistical relationships between large-scale circulation and regional climate to derive regional climate information. It is the preferred method when estimates of specific variables, especially at point locations, are the primary objective. However, this method can produce results that are not based on a true understanding of regional climate dynamics. Dynamical downscaling nests higher resolution RCMs within GCMs to generate regional climate information, an approach that allows users to incorporate topographic features such as mountains. While this method typically produces more realistic projections vis-à-vis regional

Proceedings of the First National Expert and Stakeholder Workshop on
Water Infrastructure Sustainability and Adaptation to Climate Change

22

topography, the long computational times required and the uncertainties in the models often outweigh the benefits of this method.

Recognizing the shortcomings of both methods, Casey Brown presented a related process called decision-scaling as a recommended alternative to traditional downscaling methods. Whereas downscaling focuses on increasing spatial and temporal resolutions of GCM outputs, decision-scaling uses a bottom-up approach that looks at the climate sensitivities of a system or of decisions. Climate information is then tailored to assist decision making, essentially beginning by specifying what is needed from the GCM outputs. The three steps involved in this process are (1) identifying the system or decision vulnerabilities to climate; (2) characterizing the probability of those climate hazards (risks); and (3) using systematic decision approaches to address these climate risks. Dr. Brown provided an example scenario. In the example, the decision to be made is whether it is better to build two or three dams on the Blue Nile in Ethiopia in the face of climate change. The first step after identifying this problem would be to determine if the benefit-cost ratio is sensitive to changes in climate. The second step would be to estimate the probability of changes that favor one option over the alternative. The final step is to perform the final decision analysis by multiplying the benefit-cost ratio by the probability of each scenario. Dr. Brown provided data that could be used in the hypothetical decision-scaling process to demonstrate why building three dams would be the best option, given the probabilities of each climate change scenario considered.

Click here to view Dr. Brown's presentation.

Click here to read the transcripts of Dr. Brown's remarks.

Dynamic Downscaling Efforts at EPA: Regional Linkages to NOAA and NASA Global Scale Models

Dr. Alice Gilliland, EPA ORD National Exposure Research Laboratory

The National Exposure Research Laboratory (NERL) is providing air quality modeling expertise for the EPA ORD Climate Impact on Regional Air Quality (CIRAQ) project. The objective of the project, initiated in 2002, is to examine potential climate change impacts on ozone and particulate matter using the regional-scale Community Multiscale Air Quality (CMAQ) model linked with global scale climate and chemical transport models. Through this effort, EPA has concluded increases in ozone due to climate change were fairly robust across many global-scale and downscaled climate and air quality studies. However, EPA also found that the impacts on concentrations of fine particles ($PM_{2.5}$) in the air are strongly driven by precipitation changes in the United States, but that precipitation is one of the most difficult fields to model. The models' simulations of current precipitation differ substantially from the observations. The analyses suggest a strong need for new regional climate scenarios.

NERL is currently expanding its in-house climate studies and developing new regional climate modeling (i.e., downscaling) applications. As part of this effort, NERL is relying on partnerships with other agencies using GCMs, including the National Aeronautics and Space Administration (NASA) and the National Oceanic and Atmospheric Administration (NOAA). NERL is also developing staff expertise in regional meteorological modeling and is working with National Center for Atmospheric Research (NCAR) Weather Research and Forecasting Model and others. This has led to an increased confidence in NERL's dynamical downscaling methodologies, using present-day (verifiable) scenarios. This increase in confidence will hopefully enable NERL to focus on addressing

Proceedings of the First National Expert and Stakeholder Workshop on
Water Infrastructure Sustainability and Adaptation to Climate Change

23

precipitation and other uncertainties in RCMs, where in previous work the focus on was only on air quality.

Click here to view Dr. Gilliland's presentation.

Click here to read the transcripts of Dr. Gilliland's remarks.

Web-Archive of Statistically Downscaled Climate Projections for the Contiguous United States
Levi Brekke, P.E., Bureau of Reclamation Technical Services Center

The U.S. Bureau of Reclamation in the Department of the Interior, in collaboration with researchers from Santa Clara University and the Lawrence Livermore National Laboratory, has developed a Web-based public-access archive of 112 versions of downscaled climate projections at one-eighth degree resolution (approximately 8 miles). This archive can help water managers by giving their analysts access to climate projection information at basin-relevant resolution. In addition, having an archive of a large set of these projections can help support assessments of projection uncertainty and risk-based adaptation planning by helping analysts understand the variability of different projections, which can enable them to make decisions in the face of uncertainty accordingly.

Levi Brekke provided an overview of the downscaling method used to produce the projections included in the archive. The Bureau of Reclamation used a non-dynamical, gridded method called "bias-correction spatial disaggregation." This method uses bias correction approaches at the coarse scale to adjust GCM output so that it statistically matches observed temperature and precipitation data during common historical overlap periods. It then spatially downscales the information from coarse resolution to fine, interpolating monthly temperature and precipitation to a one-eighth degree resolution.

Dr. Brekke provided an overview of the archive's Web site and its functions. The archive is available at http://gdo-dcp.ucllnl.org/downscaled_cmip3_projections/.

Click here to view Dr. Brekke's presentation.

Click here to read the transcript of Dr. Brekke's remarks.

The North American Regional Climate Change Assessment Program: A Brief Overview
Dr. Linda Mearns, University Corporation for Atmospheric Research

Since 2006, the North American Regional Climate Change Assessment Program (NARCCAP) has been working on developing high-resolution climate change simulations that investigate multiple uncertainties in regional scale projections of future climate and that generate climate change scenarios. Development of these scenarios will help in researching and assessing climate change impacts by providing critical scenario inputs to a broad community of users that includes researchers focusing on dynamical and statistical downscaling, regional analysis of NARCCAP results, and impacts studies. NARCCAP is running a set of RCMs nested in and driven by a set of atmosphere-ocean general circulation models (AOGCMs) over the conterminous United States and most of Canada.

Proceedings of the First National Expert and Stakeholder Workshop on
Water Infrastructure Sustainability and Adaptation to Climate Change

24

Phase I of the program involved 25-year simulations using NOAA's National Centers for Environmental Prediction's (NCEP)-Reanalysis boundary conditions from 1979 through 2004. Phase II involved developing climate change simulations. NARCCAP performed RCM runs with six different RCMs at 50-km resolution nested in four different current and future AOGCM runs. This was followed by conducting time-slide experiments at 50-km resolution for comparison with the RCM runs. In comparing these runs, NARCCAP is attempting to quantify uncertainty at regional scales, using probabilistic approaches. NARCCAP recognizes opportunities for "double-nesting" (i.e., conducting further downscaling of already downscaled runs at higher resolutions) over specific regions, which includes potential participation of other RCM groups (e.g., NOAA's Climate Program Office's Regional Integrated Sciences and Assessments (RISAs)). NARCCAP is making the data they collect available for the climate impacts community, the climate analysis community, and regional modelers who wish to achieve higher resolutions by double nesting. For example, these other regional modelers can take the boundary conditions from the RCM runs NARCCAP has produced at 50-km resolution and then downscale them even further over a smaller domain (e.g., the Northeast United States at 25-km resolution).

Click here to view Dr. Mearn's presentation.

Click here to read the transcript of Dr. Mearn's remarks.

Summary of Discussion Session

An initial question was broached by a climate change impacts researcher relating to the concern about the amount of time required to run the computations for an ensemble of GCMs and RCMs. The question was raised in the context of obtaining projections of changes in extreme events, such as the maximum daily precipitation in a year. While using an ensemble is time-consuming, using fewer GCMs could mean that projections are less robust. The following questions were asked: Is it possible to get away from the high resource and time costs of having to *dynamically* downscale 20+ GCM runs in order to have a usable analysis of potential precipitation at the local scale? Acknowledging that computational ability is continually improving, is there a better way to be applying computational work? The current computation is at 25-km resolution, but should the climate modeling community be doing more to achieve higher resolutions?

Regarding the difficulty of communication between water utility managers and the climate modeling community, a water utility manager commented that from the water utility perspective, the question is, "What is the decision making value of the GCM work?" For example, if there is a 10 percent decrease in precipitation, will climate models produce a level of certainty that water resource managers can take to their policy makers or ratepayers today or in the future to argue for more investment? "Actionability" is a different issue from uncertainty. At what point can we have strong enough analyses to know when to take action? A climate modeler highlighted the fact that there is a lack of agreement on what is agreed on, and that many in the climate modeling community are currently working to clarify the existence and nature of the inherent uncertainties. As for actionability, the water utility manager stated that although climate-modeling work has not necessarily been done for policy makers, it could be made more useful to them. The climate modeler responded that the concept of *actionability* needs to be reexamined.

One water research engineer highlighted the need to connect the climate modeling community with water utility managers, and identified this as a critical step in linking climate science and climate policy. Using a bottom-up approach can help strengthen this link. Such an approach could involve first looking at changes that could hypothetically present problems for the utility (e.g., a 10 percent

Proceedings of the First National Expert and Stakeholder Workshop on
Water Infrastructure Sustainability and Adaptation to Climate Change

25

decrease in precipitation or a threshold extreme rain event). Looking at the model output second can help make model outputs more useful to water utility managers within the context of system vulnerabilities.

With respect to actionability, a climate researcher asserted that while thresholds are important, one cannot defend a threshold of actionability without confidence in the outputs of the GCMs. *Statistical* downscaling methods cannot catch everything, but having an ensemble of model outputs (e.g., the Bureau of Reclamation archive) can help reach an actionable level of certainty by providing a suite of model outputs from which more robust projections can be extracted. However, where the level of actionability exists is a matter of discretion for practitioners. Much more discussion on this issue is needed to help flesh out a point of actionability vis-à-vis the level of associated uncertainty.

A climate change researcher commented that it is important not to forget that even if the downscaling methods can be shown to be effective; there is still a considerable degree of uncertainty with respect to emissions scenarios and climate sensitivities. Application of even the most effective downscaling methods, though not fruitless, still does not result in certainty, and perhaps more focus should be paid to bottom-up methods of downscaling that are comparable to the decision-scaling concept.

A water research engineer expanded on the notion of decision-scaling, explaining that in this approach there is a need to evaluate the sensitivities to a decision before looking at the decision itself. Some decisions are sensitive to climate change (e.g., increases in precipitation and temperature), while some are not. When we start with the decision, we can evaluate the various types of climate change processes that could affect the decision, and then use all tools available to get an estimate of the potential impacts in the event that those processes occur.

A water utility manager reminded the group that from the perspective of the people in his position, the uncertainties associated with climate change and climate modeling are only one type of uncertainty the water utility community deals with. Even without climate change, much of the country's underground aquifers are nearly depleted, and utilities are already working on ways to continue to meet demand in the face of these shortages. It was brought up that maybe the focus should be on looking at water shortage adaptation strategies that need to be implemented anyway, and finding synergies between these strategies and other climate change adaptation strategies, rather than focusing solely on downscaling climate change model outputs.

With respect to the notion of actionability, a water consultant noted that the actionability standard for water utility managers is typically based on past experience. The common maxim is, "I have to at least be able to handle everything I've seen in the past, and still be comfortable with my ability to handle events that exceed these limits." Operating toward the future, while knowing what has been seen and done in the past, is still the standard for water utility managers, as opposed to planning for unseen future events. Utilities need to ensure that over the long-term supplies are adequate to meet demands. Model output can be used to test the robustness of a system. If climate change means objectives may not be met, then the utilities may need to buy time. Water systems are managed for multiple objectives and different objectives should be considered. Stakeholders can be brought in to do computer simulations of alternatives.

A water resources consultant commented that from the practitioner's perspective, many water utilities are asking their engineers and consultants to develop projections for them. It would be advantageous for these practitioners and the water resources community as a whole to have a standard practice for developing these projections. Guidance is needed to ensure uniformity, and to improve decision making in the face of uncertainty, which could be a role for EPA.

Proceedings of the First National Expert and Stakeholder Workshop on Water Infrastructure Sustainability and Adaptation to Climate Change

26

5.2 Projecting Hydroclimatic Changes – Local Applications of Downscaling

A key component of adaptation is the understanding of risks that water utilities and other managers face. Typically, this has been done through estimating hydroclimatic changes at a relevant scale for water management, e.g., water basins. Outputs from climate models such as temperature and precipitation are often entered into hydrological models to estimate the change in runoff. Management models may then be used to examine implications for water resource management. This session explored how such assessments have been done and how they can be improved in the future.

Click here to read the transcript of the moderator's remarks.

Regional Modeling for the Pacific Northwest
Dr. Dennis Lettenmaier, University of Washington

In this presentation, Dennis Lettenmaier discussed the activities and findings of the Climate Impacts Group (CIG), a research group that is part of the Center for Science in the Earth System at the University of Washington. CIG's research focuses on the intersection of climate science and public policy relating to water resources, aquatic ecosystems, forests, and coasts in the Pacific Northwest. CIG is one of eight Regional Integrated Sciences and Assessments (RISA) programs funded by NOAA to research climate-sensitive issues of concern to decision makers and policy planners. CIG conducts monthly reviews of a regional climate outlook it has developed for the Pacific Northwest, and performs updates to the outlook as appropriate. The outlook is available at: http://cses.washington.edu/cig/fpt/cloutlook.shtml.

Dr. Lettenmaier presented some of the water resource-related findings from CIG's research. He explained that many water-related climate change trends are well documented, such as decreasing snowpack, but that not all changes in water resource-related processes are necessarily due to climate change. CIG's research has looked at trends in water resources in several locations, including the Yakima River basin and the Puget Sound basin, where trends show increases in winter water flows and decreases in summer water flows. Reduced summer flows are presenting problems for agriculture water uses with junior rights. Climate change-related and natural variations in precipitation and temperature are believed to be contributing to these changes in water flows, and can have indirect impacts on aquatic ecosystems. For example, fish populations in the region's rivers are sensitive to temperature changes, and many of the region's programs for protecting these fish populations have yet to begin planning for the potential temperature impacts of climate change. Another indirect impact of changes in precipitation, temperature, and stream flow is the affect on the ability to generate sufficient and reliable electricity at hydroelectric dams. Supplies for urban water uses such as for Seattle are estimated to have no or small decreases in reliability.

Dr. Lettenmaier recommended that the climate modeling community pay more attention to recognizing and addressing uncertainties in model outputs and highlighted the importance of using ensembling methods of downscaling to diminish the significance of single-model uncertainties. In conclusion, Dr. Lettenmaier drew the participants' attention to the lack of understanding between the science and user community, and identified this as an important area to focus on in future activities.

Proceedings of the First National Expert and Stakeholder Workshop on
Water Infrastructure Sustainability and Adaptation to Climate Change

27

Click here to view Dr. Lettenmaier's presentation.

Click here to read the transcript of Dr. Lettenmaier's presentation.

Hydroclimatic Modeling for Water Resources Planning in the City of New York
Dr. David Major, Columbia University

David Major provided a presentation on how hydroclimatic modeling is being incorporated into water resources planning in New York City. New York City has been accounting for climate change in its planning processes for several years, beginning with the Metro East Coast report published in 2001 as part of the National Assessment of Climate Variability and Change. Dr. Major stated that having continuity of people working on climate change has been helpful in ensuring that climate change is invariably considered in city planning processes. City agencies involved in planning processes that have or are currently incorporating climate change projections include the Department of Environmental Protection, the Metropolitan Transportation Authority, the Climate Change Adaptation Task Force, and the city's Panel on Climate Change.

The Center for Climate Systems Research (CCSR) at Columbia University is working on downscaling GCMs to develop regional climate information to assist city agencies in incorporating climate change into their plans. The CCSR has run temperature and precipitation scenarios from the Intergovernmental Panel on Climate Change's (IPCC's) Fourth Assessment Report (AR4) through sixteen GCM models, and sea-level rise scenarios from the AR4 through seven GCM models. The AR4 emissions scenarios that they have used include B1, A1B, and A2 (low, medium, and moderately high). The output from these runs has been downscaled to the New York City region from model grid boxes, and the results have been broken down into planning periods that extend from the present to the 2020s, the 2050s, and the 2080s. The results are validated using hind-casting methods.

Dr. Major presented some conclusions from CCSR's findings. Specifically, the CCSR found that over time, models project that temperatures will increase from the 2020s to the 2050s and the 2080s, but there is increasing diversity and variation as to the amount that temperature will increase. With precipitation, the models show that there will be generally more precipitation in the future, but with greater certainty with respect to the amount. For sea-level rise, the models project that there will be generally an increase in sea-level rise in future decades, but there is less certainty as to the magnitude. Melting of major ice sheets has not been included. The uncertainties in these model projections highlight the importance of erring on the side of greater change and risk. Dr. Major provided the example of leaving room to build new retaining walls if sea levels rise high enough, even if it is not yet certain that sea levels will indeed rise high enough to warrant building new walls. Dr. Major concluded by saying that climate change scenarios are now fully a part of New York City agencies' planning processes, and that the city is in the process of developing city-wide "climate protection levels."

Click here to view Dr. Major's presentation.

Click here to read the transcript of Dr. Major's remarks.

*Proceedings of the First National Expert and Stakeholder Workshop on
Water Infrastructure Sustainability and Adaptation to Climate Change*

28

Predictive Capacity in the Colorado River Basin
Brad Udall, University of Colorado – NOAA Western Water Assessment

Brad Udall presented on the findings of the Western Water Assessment at the University of Colorado, a NOAA RISA program that has been investigating flow projections for the Colorado River Basin. For years, it has been accepted that the Colorado River Basin is at risk of flow reductions. Since 1970, a temperature increase of two degrees Fahrenheit has been observed in the basin, and snowpack in March, April, and May has significantly decreased at higher altitudes. In addition, at the end of the Colorado River at Lakes Mead and Powell, water reservoir supplies and recharge capabilities are shrinking while demand increases. Several studies have sought to project the range of potential impacts of future climate change on the river's flow, but the results generated are extremely broad, with flow decrease estimates ranging from 6 percent to 45percent by 2050. The more recent studies tend to be closer to the smaller changes in runoff.

Mr. Udall explained how the Western Water Assessment is working to reconcile the discrepancies between the projections. The overall approach the program is using is to first run hydrology models during known past periods, using the same historic data in each of the different hydrology models, to produce changes in streamflow with respect to temperature and precipitation. The second step is to drive the hydrology models with future climate change projections. This process has resulted in several major findings. With respect to precipitation, the models showed that a 10 percent change in precipitation produces an approximately 20 percent change in streamflow. Temperature was more variable, however, with a one degree Celsius change in temperature producing changes in streamflow ranging from 2 percent to 9 percent depending on the model.

Dr. Rajagopalan Balaji and colleagues at the University of Colorado recalculated Barnett's study on drying of the Colorado River[1] (which found a 50 percent chance of Lakes Mead and Powell going dry by 2021) and found a 5 to 12 percent chance of Lakes Mead and Powell drying by 2026.

Mr. Udall noted that the Brookings Institute published a paper by Pat Mulroy calling for a single Los Alamos type entity to combine current and potential future research on water resources and for more collaboration among water agencies.

The results of the project's efforts highlight the importance of modeling at high areal resolutions. In addition, Mr. Udall noted that a future emphasis will need to be on utilizing consistent GCM-derived temperature and precipitation projections to drive different hydrology models, so that outputs can be compared.

Click here to view Mr. Udall's presentation.

Click here to read the transcript of Mr. Udall's remarks.

---

[1] Barnett, T. P. and Pierce, D. W., 2008: When will Lake Mead run dry? *J. Water Resources Research*, v. 44, W03201, doi:10.1029/2007WR006704.

Summary of Discussion Session

To open the discussion session, a climatologist raised the question, "What would happen to water resource managers' "actionability" thresholds if we experience little or no additional climate change impacts over the next 10 years, but still understand that changes are inevitable in the long-term?" One response from a climate change impacts researcher was that investing in local infrastructure requires that local planners look more than twenty years into the future anyway, so it would not be very problematic if we do not see significant climate change impacts in the next ten years. Many local governments, particularly Seattle, New York City, and Boston, have been planning for climate change for several years in the face of current climate change impacts, which have not been overwhelming. Maybe EPA could adopt rules to require local governments to incorporate climate change into their plans to ensure that they do so even if significant climate change impacts are not experienced in the near future.

A water utility manager made the comment that if the public does not see evidence of climate change in the near future, it could be difficult for water utilities to ask ratepayers for additional money to implement climate change adaptation strategies. A climatologist stated that there is a 10 percent to 15 percent chance that there will be cooling effects over the next decade before temperatures begin to increase in subsequent decades. The challenge will be explaining the importance of adaptive measures in the immediate future to the public. How do we make the long-term, global impacts of climate change tangible on the local level in the near future? A hydrologist and climate change impacts researcher made the point that the impacts of climate change are already evident in some places in the country (e.g., places with decreasing snowpack), and that the public is already sensitive to the need for adaptive measures.

A climate researcher raised the issue of discrepancies in studies on the impacts of climate change. Pointing to Mr. Udall's presentation on predicting streamflow in the Colorado River basin, the commenter made the point that researchers are taking many different approaches to exploring the same problems, and that we have no effective way to verify the results, which are often inconsistent. A climate change impacts researcher agreed, noting that while most studies agree on the direction that trends will take (e.g., a decrease in streamflow), there is rarely concordance on the magnitude of the change.

A climate change researcher raised the question, "What level of precision is necessary for policy makers in order for them to act?" If we know the direction of the change, is that sufficient? Or is knowledge of the magnitude of the change absolutely necessary? A water utility manager responded that from the utility's perspective, they are accustomed to dealing with variability and uncertainty. In addition, they have little room for error and must always judge the effectiveness of their plans on the ability of the plans to succeed in extreme situations, even though the ratepayers do not always recognize the amount of money that is spent planning for events that rarely, if ever, occur. This sort of thinking applies in the case of climate change. It is hard to know what climate change impacts will do to water resources and utility operations. Moreover, we often overlook indirect effects on other key issues. For example, reduced streamflow in the Colorado River basin could result in an inability to generate electricity at Hoover Dam, which could have significant implications across a large region. Water resource managers and utilities need to continue to plan for low probability, high impact events. We do not wait for absolute certainty; rather, we act with a preponderance of evidence.

Another water utility manager noted that the more intense stormwater events is another indirect effect of climate change impacts that many wastewater utilities will need to prepare for. Some have

Proceedings of the First National Expert and Stakeholder Workshop on
Water Infrastructure Sustainability and Adaptation to Climate Change

30

started to incorporate stormwater contingencies into their plans, but much more progress is needed. A wastewater manager made the comment that the science community is not yet providing the practitioners with adequate information to plan for more intense stormwater events. The point was also made by another wastewater manager that wastewater treatment facility managers are in a different situation from water utility managers in that the public perceives them differently.

A water resources manager asked what information the wastewater treatment community needs in order to begin incorporating climate change into their planning documents. A climate change impacts researcher suggested that less variability in data about intense stormwater would be very helpful. In New York City, the city has been planning wastewater treatment infrastructure using intensity-duration-frequency (IDF) curves, but there is concern that developing IDF curves for the future could lead to lawsuits. A climate change impacts researcher commented that in King County, the county's wastewater treatment department has been looking at downscaled climate information, but not as far into the future as the water resource managers. The point was also made that the wastewater treatment community has to deal with additional concerns, namely environmental standards (e.g., with stormwater ponds), that are not applicable to the water resource management community.

A water resources engineer raised the issue of whether modeling approaches can be revised to be more effective. Engineers would benefit from having access to models that produce output based on the integration of design storms, which would enhance long-term planning. A research hydrologist noted that there is a significant need for the science community to provide information on design storms.

A hydrologist brought up the issue of the need for up-to-date data. Many people are concerned only about the climate change projections and variability that comes out of models, but much of the data that go into calculations of intense precipitation is obsolete, sometimes dating back to the 1960s. Federal Emergency Management Agency (FEMA) flood maps are an example of how model outputs are used even though the inputs are no longer appropriate. It seems that this issue is of little concern among the practitioner community compared to the variability of the outputs. There has been very little funding provided to NOAA to update these data.

A hydrologist and climate change impacts researcher commented that it is important to understand where climate sensitivities reside. Downscaling methods often produce results that could be plausible, but are not suitable for the short-term because they often require too much computation time. For the short-term, 25-hour storm analyses require long computations and bias correction before the information can be used, and this requires time. At what scale is it necessary to model impacts in the face of short-term needs? How far down must we scale climate information in order to be able to take action?

A hydrologist raised the issue of how the public will react to climate change, stating that adaptation is a behavioral-social issue and that we must treat it as such if we are going to take action. It is extremely important that the water resources community have at its disposal the tools necessary to communicate to the public the importance of adaptation. These tools exist but they are not standardized and there is no accepted process for training those who would be using the tools. To effectively incorporate climate change into water resource management, it is necessary for these tools to be publicly available to the water resource management community, and standardization and education in using these tools is imperative. A hydrologist responded that these tools, while extant, are outdated. EPA should encourage the teaching of 21st century tools, such as on probability and risk, and should require regulators to use the new tools. A participant from the

Proceedings of the First National Expert and Stakeholder Workshop on
Water Infrastructure Sustainability and Adaptation to Climate Change

31

modeling community stated that it is absolutely imperative that the water management community prioritize efforts to develop standardized tools.

A climate change researcher raised the question, "What form should such a tool take? Should it be something comparable to the 7Q10, the seven-day, consecutive low flow with a ten-year return frequency (the lowest stream flow for seven consecutive days that would be expected to occur once in ten years)?" A hydrologist said that the 7Q10 is updated every few decades, but it depends on agreements with state agencies.

A climate change researcher suggested that EPA should maybe not focus so much on GCM and regional climate model (RCM) outputs too far into the future, and should adopt a "segmentation" approach that considers immediate short-term impacts that need to be addressed first. Hundred-year floodplains might not be as critical as the once-a-year flood, or the one-year-in-ten flood. We need to break down the time periods to identify different decision points.

A hydrologist made the point that from the water resource manager's perspective, when looking at future projections, it is important to start by considering one's objectives (i.e., what one is managing for), which are often complex and interrelated. Water managers do not manage for the environment as a whole; they manage for a suite of environmental needs. Another water manager noted that this is the difference between top-down and bottom-up approaches to downscaling, and that it seems like bottom-up methods might be more appropriate given the diversity of objectives. A climate change impacts researcher noted that an ensembled top-down approach can also be effective in a given situation. Some objectives are more conducive to top-down approaches (e.g., hydropower production), while others are not (e.g., ecosystem management).

A hydrologist noted the importance of recognizing observed data when developing and modifying models. A great deal of attention is paid to the models themselves, and the outputs they generate, but we need to pay attention to whether the models reflect reality. We should also be concentrating on archiving observed data in a useable fashion. A climate change impacts researcher noted that while it is good that the data be archived and made available, it needs to be collected and presented in a helpful way, possibly in a single location. A climatologist said that stream gauges and the hydro-monitoring network are the backbone of climate observations and that much of the country's data monitoring infrastructure is being shut down or not replaced due to lack of funds. A water resources consultant responded that there needs to be a Congressional mandate requiring EPA to regulate for the best available information, and to invest in enhanced observation capabilities.

5.3 Evaluating Hydoclimatic Change for Water Infrastructure Adaptation – Part I

How to use observations is a critical part of adapting to climate change. This session explored what hydrologic changes are and are not being detected in the observed climate record. The detection of trends is a matter of interpretation and, as pointed out, made more difficult when the monitoring network is degrading. The challenge of adapting to climate change is complicated when observations are inconsistent with model projections.

Click here to read the remarks of the moderator (Dr. Dan Sheer).

Proceedings of the First National Expert and Stakeholder Workshop on
Water Infrastructure Sustainability and Adaptation to Climate Change

32

Hydrology and Climate Change: What Do We Actually Know?

Dr. Robert Hirsch, U.S. Geological Society

Robert Hirsch presented on the topic of water resource planning in the post-stationarity age. He began with an overview of the basics of water resource planning, explaining that planning is centered on risk-cost tradeoffs. These tradeoffs have been the subject of much research and information collection, beginning with the Harvard Water Program in the late 1950s. Dr. Hirsch also highlighted the fact that water planning analyses require planners to make certain assumptions about the distribution of hydrologic variables (e.g., streamflow), which makes it necessary to reevaluate current practices that have been in use since the 1950s, which are based on an assumption of a stationary climate (i.e., observed data from recent decades can be used to project future climate conditions).

Recent studies suggest that the concept of stationarity is no longer appropriate for water resource planning and that finding a successor to this approach is a critical step in adapting to climate change. Moreover, modeling should be used to synthesize observations – not replace them – and in a nonstationary world, continuous and accurate recording of observations is critical.

There is currently significant variability in the precision of data on streamflow conditions. For instance, data show that flow timing generally shifts in areas where snowpack is significant, even though there is not always evidence of runoff volume changes. The data also show that low flows and average flows are predominantly increasing, while changes in flooding and changes in groundwater remain very unclear. With respect to flooding, the data do not provide clear evidence that flooding is getting worse, only that flood damages are increasing. The latter may be the result of changes in society (e.g., increasing construction in floodplains and property values). As for changes in groundwater, Dr. Hirsch explained that the data provide very little clarity about what can be expected in the future.

Dr. Hirsch presented several graphics depicting changes in stream flows in various locations around the country, explaining what historical data can and cannot tell us about the future. One important thing the data do tell us, Dr. Hirsch pointed out, is that despite public beliefs that climate change has been – and will continue to – make the globe hotter and drier, evidence shows that approximately 50 percent of locations around the nation experienced increases in annual median flows from 1941-1971 to 1971-1999 and almost no sites show a decrease. In addition, changes in flows have varied considerably depending on the location of the stream. As an example, Dr. Hirsch showed the discrepancy between flow changes between locations in North Dakota and Iowa (where floods have increased over the past 150 years) and locations in Georgia and Utah (where floods have decreased).

Dr. Hirsch outlined a nonstationary alternative approach to water resource planning that involves two key concepts. The first is to pay attention to what is actually happening hydrologically since climate models do not provide reliable answers. For example, it is important to expect quasi-period phenomena that climate science cannot yet explain (e.g., El Niño), and these limitations need to be acknowledged and addressed appropriately. The second concept is that we should not lose track of other major change drivers (e.g., groundwater depletion, eco-flow requirements, nutrient enrichment, and demographic/economic/energy demands). To highlight the importance of keeping track of multiple change drivers, Dr. Hirsch provided an example that compared greenhouse gas concentration in the atmosphere and nitrate in rivers and aquifers – two global continental scale environmental changes with implications for water resources. Dr. Hirsch presented several graphics

Proceedings of the First National Expert and Stakeholder Workshop on
Water Infrastructure Sustainability and Adaptation to Climate Change

33

on how each of these global changes can influence water resources, and explained the sort of data that are available to evaluate these impacts.

One of the overarching points of this comparison was to underscore the importance of accurate measurements. Dr. Hirsch concluded by explaining the extreme consequences of failing to continuously and accurately record observations, informing the group of the high rate at which stream gauges are currently being shut down (including 100 gauges in 2007 alone).

Click here to view Dr. Hirsch's presentation.

Click here to read the transcript of Dr. Hirsch's remarks.

Precipitation Frequency Atlas of the United States: Update and Issues
Geoffrey Bonnin, National Oceanic and Atmospheric Administration, National Weather Service

Geoffrey Bonnin presented on the topic of current activities within the National Oceanic and Atmospheric Administration (NOAA) to update the NOAA Precipitation-Frequency Atlas (Atlas 14), a principle standard for national rainfall intensity and frequency estimation which is based on data that – in some places – date back to 1962. Atlas 14 is accessible online through NOAA's Precipitation Frequency Data Server, which provides interactive tables and charts with geographic information systems (GIS) maps. Users can vary outputs for seasonality and temporal distribution. The server is accessible at http://www.nws.noaa.gov/ohd/hdsc.

Funding for observations has virtually disappeared in the last decades. Departments of transportation in state governments are the major source of funding.

Mr. Bonnin reviewed some of the observed changes in intense precipitation. The 100-year, 24-hour values show a mix of increases and decreases in the Southwest. In the East, mid-West, and mid-Atlantic, there is more of a tendency for increases than decreases.

NOAA's current updates to Atlas 14 have resulted in a number of changes (e.g., changes in 100-year, 24-hour precipitation measurements). Mr. Bonnin explained that these changes are due to improvements in data and enhanced statistical and spatial interpolation techniques. He also made it clear to the group that NOAA does not think that all of these changes are indicators of climate change. Based on observations, the impacts of climate change on precipitation frequency have been small (compared with the error in estimation). Models show that projected changes over the next 50 years are small with respect to the errors associated with the estimates of precipitation frequency. At 100 years, the projected changes are considerably larger, suggesting that it may be more appropriate to consider 100-year frequency projections rather than focusing on 50-year frequency projections. These results indicate large differences between models and between forcings. The fact that the model results are only applicable down to 200-kilometer resolution highlights the fact that the downscaling approach is questionable.

In continuing activities, Mr. Bonnin explained that NOAA will use unadjusted historical data because there is currently little understanding of how the data should be adjusted.

Click here to view Mr. Bonnin's presentation.

Click here to read the transcript of Mr. Bonnin's remarks.

Proceedings of the First National Expert and Stakeholder Workshop on
Water Infrastructure Sustainability and Adaptation to Climate Change

34

National Hydroclimatic Change and Infrastructure Assessment: Region-Specific Adaptation Factors

Dr. Y. Jeffrey Yang, P.E., U.S. EPA National Risk Management Research Laboratory

In this presentation, Jeffrey Yang described the efforts of the EPA Water Resource Adaptation Program (WRAP) and explained how the program's investigations contribute to research on approaches to water resource adaptation. WRAP's objective is to provide data, tools, and engineering solutions for adaptation to climate, land use, and socioeconomic changes. In this new program, scientists and engineers investigate potential effects of climate change on the nation's watersheds and water infrastructure. Based on the results of these investigations, practical and effective adaptation solutions are being developed. The research approach has three basic elements: (1) investigating hydrologic effects of climatic change and defining the water resource needs of future socioeconomic conditions using tools such as climate modeling, robust statistical analysis, and water availability forecasting; (2) developing adaptation methods, primarily focused on advanced and innovative engineering techniques and solutions; and (3) developing and providing end users with tools for water resource adaptation. This information is used to establish adaptation measures for specific regions and watershed basins (e.g., wastewater reuse in water-stressed Florida, the Great Plains, the Southwest, California, and other West Coast regions). More information on the program is available at: http://www.epa.gov/nrmrl/wswrd/wqm/wrap/.

Dr. Yang provided an overview of the current paradigm in infrastructure and water program adaptation, explaining the significance of uncertainty – in terms of engineering options and costs – and timeframes with respect to different rates of climate change. He showed how these uncertainties over different timeframes can impede effective decision making for infrastructure planners.

WRAP has adopted a systematic approach to assessing hydroclimatic changes and impacts that recognizes inherent errors and uncertainties in model outputs. This approach considers several essential questions, including whether hydroclimatic changes are tangible; whether the scientific evidence of these changes is sufficient to warrant action, and if so, what are the changes within infrastructure and water program planning horizons; and how to adapt.

Using this approach, WRAP is currently conducting a Nationwide Hydroclimatic Changes and Adaptability Assessment. This assessment looks at hydroclimatic changes, land use changes, socioeconomic developments, and infrastructure conditions and integrates temporal periodicity, spatial correlations, and hydroclimatic province classification. The result is a regional-level overview of expected changes due to climate change (e.g., sea-level rise along the Atlantic, Pacific, and Gulf Coasts) and lists of specific potential adaptation factors in these areas (e.g., population change and its relevance for changes in sea-level rise, storm surge, and disruptive meteorological events). These adaptation factors illustrate large regional differences in hydroclimatic and land use changes and show that primary adaptation factors are often region-specific. These results will enable water resource planners to understand key vulnerabilities in their region and to plan accordingly.

In conclusion, Dr. Yang identified the need for further refinement of regional impacts and geographic distributions, as well as a need for an examination of adaptive integrated water resources management through a prism of risk management that accounts for rates of change, different planning time horizons, uncertainty, energy demand and supply, and economics.

Proceedings of the First National Expert and Stakeholder Workshop on
Water Infrastructure Sustainability and Adaptation to Climate Change

35

Click here to view Dr. Yang's presentation.

Click here to read the transcript of Dr. Yang's remarks.

Summary of Discussion Session

The discussion was initiated with a comment by a hydrologist who stated that rules evolve when either they stop working or it becomes evident that implementing them costs too much. The kind of information that was presented by Mr. Bonnin is the type that drives rule evolution. The rules are not being updated and that may be because they are working. It may cost too much to change the rules and climate change may not be enough of a reason to change them. While there is not much information on climate change that leads us directly to the ability to plan specific designs, there is no reason why we should not anticipate climate change to avoid larger problems in the future through such measures as putting in footings to adapt to sea-level rise.

A water engineering researcher pointed out that there are limitations to what climate science can explain and that it is good to make a distinction between what the models can do and what climate science can do. In most cases, assumptions must be made about the future. With climate science, we can make the best predictions for the future using nonstationary approaches. If the climate community can come up with a substitute for stationarity, what is the potential for putting this new regime into practice? Are the prospects realistic if the research community is focused on more narrow topics? A research hydrologist responded that this will require a funding source that is not interested in narrow results, and that it is not clear if any federal agency is interested in this at the moment. The research community should be looking for these sorts of opportunities. Maybe industry can play a role, but the motivation needs to come from the federal government. Moreover, we cannot be focused on achieving instant results because there needs to be a long-term investment.

A hydrologist commented that when NOAA was first publishing its Atlas 14 updates, it was expecting a large volume of lawsuits from developers saying that their environmental compliance expenses would be increasing significantly as a result of the changes in Atlas 14. Surprisingly, there were no lawsuits. The assumption is that practitioners are using the updated data without paying as much attention to confidence intervals. This suggests that maybe NOAA can just add uncertainty factors into tables and people will pick it up and use it. But there are issues when it comes to regulation (i.e., who would regulate?). Virtually every county has its own design manuals. The NOAA Atlas publications are supposed to be federal standards, but there are few in the federal community who pay attention to them because the federal community focuses on regulating and funding. Perhaps the federal community should be active in saying things like "thou shalt use this standard in all designs."

Another hydrologist said that as for changing hydrologic practice, practitioners say "We are aware of climate change processes, but how do we use this information to make estimates of 100-year floods?" His reaction is to view the problem from two angles. When you look at the empirical data, you need to ask whether there is a strong trend – it is hard to see trends in the upper tails of data. When you look at the models, you need to ask whether they work. The Intergovernmental Panel on Climate Change (IPCC) tells us that models do not predict extremes well. These two avenues suggest that there is not enough knowledge on which to base changes in practice. Moreover, there are other areas to consider. For example, urbanization has a huge role on the magnitude of floods, due to the impacts of impervious surfaces. Another is the effect of large dams: there is no empirical evidence as how dams affect floods. We should be incorporating these areas into integrated planning.

A water researcher commented on the complexity of the issue and suggested focusing on two questions: How should we look at historical data, and how should we incorporate these data into models? We need to think about timeframes that are relevant to all the hydrologic and climate modeling communities involved (including engineers and water managers). If you look at the ranges of timeframes, many engineers are not thinking about the uncertainty embedded in the data that are available, but it is necessary to incorporate uncertainty. Standards for practice will need to incorporate uncertainty in design management.

Returning to the initial comment of the discussion, a climate modeler mentioned that in the context of climate modeling results, current results are too uncertain to be used in decision protocols that already exist. So the questions are, Are there thresholds of uncertainty, and Can we – given a level of uncertainty – have a target level of uncertainty to reduce it to in order to reach a point where we are comfortable taking action? Moreover, how do we get the two sides (i.e., the climate modeling community and the water resource planners) together to address these questions? Whether and to what degree it is appropriate to be confident in climate model outputs is an important issue. As was done for the IPCC report, it is important to combine information into ensembles. To do this, the hydrologic and climate modeling communities need to work together.

A climate change researcher made a point that we are talking about long-term investments and decision making, which leads to the question of what kinds of information are being observed, and what level of detail in these observations is actionable for users? A key question to be answered is, "Where do we direct research to improve decision making?"

A research hydrologist said that the hydrologic and climate modeling communities can work together, in fact, they need to work together. The commenter made the point that the USGS is committed to being involved in the climate modeling community's efforts because it is critical to have people with hydrology-based perspectives involved in key modeling activities. Unfortunately, the overlap between the two communities is still very small at this point. A hydrologist commented that he does not see splits between the two groups today. We need more research and answers to key questions, but there are not enough people working on these issues right now. Despite this, we can be looking to the collective community of engineers, designers, and planners that are using federal rules and standards that the hydrologic and climate modeling communities publish. Part of the reason why we publish confidence information is to help these practitioners. It is important to make the information more probabilistic to help them make informed decisions with uncertainty ingrained. In addition, the people on the ground need to be approached with the issues we are talking about today, but in a less esoteric way. We need to give them rules and standards that are understandable and that they can put into practice.

A water researcher mentioned that EPA is currently working to incorporate probabilistic data into its hydrologic modeling through the WRAP program. In addition, there is currently an initiative on the part of engineers working to incorporate hydrology into modeling. A strong relationship with engineers can be a key asset for climate and hydrologic modelers in their efforts relating to climate change.

A water research engineer commented that he and fellow engineers always seek to take pragmatic approaches to problems and to use tools at their disposal. He posed the question to practitioner engineers, "What tools do you need?" A water engineer responded that he will use data available to him, but has concerns about models. Specifically, he commented that models are generally best when based on historical data. Precipitation and streamflow have direct relationships to many of our hydrologic indices, but periodicities are not being used in the models – some rule curves ignore them. Information on paleo-climates is not available. Where we have changed rule curves we have had increased efficiencies, but they have been ignored. The opportunities are there to use this next 10-50 year period, while we are settling down the models, to look at observational databases. At the same time, current models do have value (e.g., some have been used to predict floods and storm surges). Speaking on behalf of the practitioner community, practitioners do think outside the box. We need to focus on the trends in observations and bring these into our models. Many engineers are implementing observational data, but they are not being incorporated into the models. Overall, there is a lot of opportunity for climate modelers, engineers, and water managers, as well as for students.

A water utility manager mentioned that he is interested in consumers' points of view. Without disputing that the data are what we can best base action on, as a water policy maker trying to plan for unprecedented change, we have to do what we can to understand what the future holds. If historical data and observations do not give us a good answer, and modeling does not give us what we are looking for, we need a hybrid approach that we can use to make our decisions. The answer to the question, "What we are looking for?," is probabilities and risk assessment. The question is therefore, "Are we going to get probabilities and risk factors out of climate models on the parameters we care about (e.g., temperature, precipitation frequencies, and storm intensity)?" Will this give us an understanding of what the next 100 years will look like, and move us toward actionability? Or is 100 years too far into the future for us to plan? Without probabilities and risk factors, we cannot go to our ratepayers for more money to invest in infrastructure.

A hydrologist commented that we are trying to figure out how to put trends in climate change into engineering practice. This consists of working with rules that are based on observations (e.g., 100 year events, confidence intervals), since these are parameters that we use to develop the basis for engineering rules. Information that comes out of climate change models now informs our judgment but does not inform the rules. If we are going to get model output into the rules, we need to demonstrate that the current rules do not work or are too expensive. To do so, we need data and sampling. We currently do not sample, so we need to put up more sampling stations. The commenter suggests that the National Environmental Policy Act (NEPA) could be used as a model for sampling to figure out how much it costs to make things work. Additionally, we could set up requirements to do evaluations at higher intervals, so that people can look at the evaluations to get better information.

A water researcher made the point that when we look at large uncertainties and long timeframes, as we customarily do for water infrastructure investment, there is a need to evaluate plans. Once you have evaluated and reevaluated these plans, you can incorporate observed data to give you confidence to reduce the uncertainty in the climate models. There is a need to do this type of assessment, and to incorporate it into the studies and actions.

A water resources consultant commented that many practitioners agree that we need to move to a risk management approach that accounts for uncertainty. There is a need to do this using alternatives to the regulatory approach. For areas not covered by regulations, we need to incorporate robustness and uncertainty. Uncertainties should include broad probability distributions, for example, including evapotranspiration as well as precipitation.

Proceedings of the First National Expert and Stakeholder Workshop on
Water Infrastructure Sustainability and Adaptation to Climate Change

38

5.4   Evaluating Hydoclimatic Change for Water Infrastructure Adaptation – Part II

A number of water utilities have been moving ahead to understand their vulnerabilities to climate change and make adaptation decisions. This session explored how those utilities have been assessing vulnerabilities and making decisions on adaptation. In particular, this session focused on how utilities have used climate model output and other information on climate change to aid them in decision making.

Click here to read the remarks of the moderator (Dave Behar, San Francisco Public Utilities Commission / Water Utilities Climate Alliance).

Climate Vulnerability Assessments
David Yates, National Center for Atmospheric Research

The National Center for Atmospheric Research (NCAR) is partnering with the Water Research Foundation (WRF; formerly the American Water Works Association Research Foundation) to work with water utilities to develop decision tools to facilitate assessments of water utility vulnerabilities to climate change and adaptation options. This work builds off a past project between these two partners that resulted in the production of a primer on climate change for the drinking water industry. More information on this collaborative can be found at http://www.isse.ucar.edu/awwarf/.

David Yates began his presentation by offering an overview of the difficulties of using climate model outputs from the perspective of the water utility. He explained how most general circulation models (GCMs) have between 1 and 2 million calculation points, but even with this many points the models do not produce enough certainty in areal resolution. These models produce temperature projections with relatively higher-certainty than precipitation projections. The model grids are too large, however, to be useful for utilities, and the models often overload computers even at resolutions of 100 kilometers, where 10 kilometer resolution would be ideal. To bring model output down to a level usable to utilities, the community uses what is called "parameterization" (or GCM "drizzle"), but strategies for doing so are still primitive, and the parameterized models cannot tell us much more than "When it rains, it is going to rain a little harder."

Dr. Yates next provided the group with an overview of current activities of the NCAR-WRF collaborative effort that are aimed at resolving some of these model downscaling issues. Specifically, they are working with utility partners to develop structured processes for explicitly considering climate change in water utility decision making. This decision analysis process involves four steps: problem structuring, deterministic analysis, uncertainty analysis, and evaluation of alternatives. Problem structuring involves looking at the utility's goals, alternatives, information, and values before conducting a deterministic analysis, which consists of developing a model for the decision to be made and performing a sensitivity analysis to identify key variables. The uncertainty analysis step requires that utilities represent key variables with probabilities and determine the best plan under uncertainty. After evaluating each of the alternatives, utilities can repeat the process in multiple iterations to produce more robust results.

Dr. Yates described several example problems that NCAR and WRF's utility partners are concerned about and explained how NCAR, WRF, and various other consultants and research groups are working together to develop tools to support these utilities' decision making efforts. He explained how the collaborative partnership is working with utilities to implement planning strategies that are not based exclusively on historical data and observations, but are based on models. Dr. Yates drew comparisons between critical questions involved in hydrological modeling and planning modeling, noting that there is a need for coordination between the two. He explained that NCAR is working with other organizations, including EPA, to develop the Water Evaluation and Planning System (WEAP), an integrated resource management tool that combines hydrology and water planning models. Dr. Yates concluded by providing an overview of how this tool has been used by two of its partner utilities in their decision making processes to provide them with simulations of water demands and supplies based on user-created variables and user-managed scenarios. More information on this tool is available at http://www.weap21.org/.

Click here to view Dr. Yates' presentation.

Click here to read the transcript of Dr. Yates' presentation.

Strategies for Assessing Impacts and Adapting to Climate Change for Wastewater Utilities
Laura Wharton, King County Department of Natural Resources and Parks

In response to an executive order issued in 2006, King County developed a Climate Change Plan that summarized climate change projections in the Pacific Northwest, based on research by the University of Washington's Climate Impacts Group (CIG) and identified action items for the county to take to adapt to projected changes. The plan included action items for the wastewater treatment division related to reclaimed water and strategies for the division to manage wet weather impacts of climate change. The division owns and operates a wastewater collection and treatment system that serves a population of 1.4 million people, collecting wastewater from 34 local sewer agencies and conveying it to two treatment plants (a new treatment plant is currently under construction).

According to the CIG's projections, the region can expect higher temperatures by 2100 (increases of roughly 1.8 degrees Fahrenheit every 25 years); potential increases in precipitation with more rain than snow likely; snowpack decline with earlier runoff; increased risk of floods and drought; rise in sea levels; and potential impacts to groundwater. Of these impacts, the wastewater treatment division is focusing primarily on threats from rising sea levels, intense storms, and increased capacity needs. Approximately 40 structures, including one of the treatment plants and numerous pipeline systems are periodically affected by tides and storm surges from the Puget Sound. The division used the research supplied by the CIG to evaluate its vulnerabilities to climate change and found that sea-level rise could lead to impacts on the division's facilities by 2050, a time span that is approximately equal to the expected life of a new treatment facility, which underscores the importance of immediately incorporating climate change into planning.

Laura Wharton described current actions by the division to implement the action items outlined in the Climate Change Plan, including promoting regional water supply resilience by maximizing development and use of reclaimed water from the wastewater treatment system and exploring reuse approaches. She provided an overview of how treated wastewater can be reclaimed and reused for a variety of uses and described a number of benefits from reclaiming water. Ms. Wharton described the division's activities in response to the plan's action items. These activities involve continuing the division's existing reclaimed water programs (currently 300 million gallons per day are reclaimed for irrigation and cooling uses), identifying customers for reclaimed water from the new Brightwater Treatment Plant, and completing a draft reclaimed water comprehensive plan by 2011. This new comprehensive plan will investigate the conditions under which future reclaimed water

Proceedings of the First National Expert and Stakeholder Workshop on
Water Infrastructure Sustainability and Adaptation to Climate Change

40

investments should be made, opportunities for reclaimed water use, and regional issues and implications for develop future projects.

Ms. Wharton introduced another component of the county's Climate Change Plan, the Adaptation and Planning Response Vulnerable Facilities Inventory, which was developed for county divisions that are responsible for long-term asset management to help them understand potential risks to their assets as a result of climate change. Using this inventory, the division is developing strategies to manage wet weather impacts of climate change to the sewer system. This involves identifying facilities that might be impacted by storm surge and sea-level rise, and then developing a methodology to combine these projections to analyze impact thresholds and characterize impacts. The inventory work is also helping the division identify adaptive strategies for affected facilities and include findings in comprehensive plans and facility design. Ms. Wharton provided an overview of the sources of data that the division is using to provide input into the inventory (e.g., sea-level rise forecast data from CIG) and presented a sample output from the inventory that illustrates the sea-level rise vulnerabilities under multiple sea-level rise scenarios (ranging from low sea-level rise to rapid ice sheet melt) for a specific facility. From this exercise, the division has discovered multiple instances where plans will need to be revised to address vulnerabilities (e.g., the design for the sampling facility and flow monitor vault sites at the Brightwater Treatment Plant was raised by about five feet). The division has also decided to conduct an analysis of sea-level rise impacts on system hydraulics and to include sea-level rise as a planning factor in all future projects. In addition, the division plans to review sea-level rise literature every five years and address changes in five-year updates to the conveyance system plan.

In conclusion, Ms. Wharton explained that the process the division is currently engaged in is a dynamic one that will require clear and continuous communication with the public. In addition, the division will need to focus on evaluating investment risks and determine appropriate thresholds to spur action and expenditures. New approaches the division is considering include building in resilience to changes, promoting and funding sustainability, and integrating processes to improve results (e.g., using reclaimed water as a resource).

Click here to view Ms. Wharton's presentation.

Click here to read the transcript of Ms. Wharton's remarks.

Implicit Climate Change Adaptation: Modifying System Operations for Turbidity Control
Paul Rush, New York City Bureau of Water Supply, and Dr. Daniel Sheer, HydroLogics, Inc.

The New York City Department of Environmental Protection (NYCDEP) operates a water supply system that delivers approximately 1.2 billion gallons of water daily to nearly half the state's population, and about 90 percent of this supply comes from the Catskill/Delaware system west of the Hudson River. In 2007, the EPA granted the department a waiver through 2017 that permits it to use this water supply without having to filter it. Maintaining filtration avoidance requires diligent control of source water turbidity, which refers to the cloudiness of a fluid due to suspended solids such as naturally occurring silt (e.g., due to erosion), since high turbidity can have potential impacts on the disinfection process. Many GCMs under a variety of Intergovernmental Panel on Climate Change (IPCC) emission scenarios suggest that the region will experience increased precipitation and more frequent intense precipitation events in the future, which can lead to increased turbidity.

Proceedings of the First National Expert and Stakeholder Workshop on
Water Infrastructure Sustainability and Adaptation to Climate Change

41

NYCDEP has conducted studies to help it understand the sources of turbidity in the Catskill system and to help inform structural and nonstructural strategies to prevent, manage, and control turbidity. The work was completed by linking a water supply system mass-balance model called OASIS and a reservoir water quality model called W2. Using these models in tandem, NYCDEP evaluated six alterative turbidity control options (e.g., dividing weir crest gates and building basin diversion walls) for the terminal reservoir in the Catskill system, including five infrastructure development alternatives and one operations-based alternative. For each alternative, NYCDEP calculated the expected frequency of days on which alum treatment would be required. The studies revealed that nonstructural system operational changes alone can significantly reduce the turbidity of the terminal reservoir of the system; when combined with infrastructure alternatives, the results suggest that the number of alum application days could be reduced to near zero. These results are contrary to the earlier belief that major infrastructure investment would be required to control turbidity.

While the combined OASIS-W2 model was useful for conducting these studies, the studies concluded that the development and implementation of a real-time system Operation Support Tool (OST), which combines water quality and water supply data with forecast inputs, along with the construction of selected infrastructure improvements is the most cost-effective means to achieving turbidity control. In addition to the water supply benefits, the implementation of OST will provide a better understanding of water supply risks associated with operational changes, and will allow NYCDEP to react to changing conditions in real time when considering system water quality, downstream flood events, and cold water fisheries habitats.

In conclusion, Paul Rush identified several research and development needs, including needs for new forecasting tools (e.g., El Nino Southern Oscillation (ENSO)-based ensemble inflow and demand forecasts) and development of tools to assess the impacts of land use changes, groundwater pumping, and climate change on runoff, water demand, and water quality. In addition, more research needs to be focused on impacts assessments of operations modifications on the reliability of physical facilities, and on assessments of legal liability issues surrounding adaptation of water supply operations to climate change (i.e., if we are helping neighboring water systems or fellow agencies with other environmental objectives, are we putting ourselves at risk of liability suits?). Mr. Rush also highlighted the importance of recognizing that building new infrastructure should not be the default solution to managing water supplies; rather, the focus should be on how the infrastructure is used. Utilities need to invest in analytical tools to ensure that the operational capabilities of existing infrastructure are maximized and tools for doing so need to be able to be used to meet consumer and stakeholder expectations in the face of uncertain conditions.

Click here to view Mr. Rush's presentation.

Click here to read the transcript of Mr. Rush's remarks.

Summary of Discussion Session

To open the discussion session, a water utility manager asked how critical it is to improve predictive tools. A local water resource manager responded that it is very important, since it can help water utilities and water resource managers make better decisions and provide better benefits to their customers.

A water resources manager commented that one of the things that has been useful to local governments is to have access to the climate modeling research of state universities because the public notices when their findings are published. This takes the pressure off local governments so

Proceedings of the First National Expert and Stakeholder Workshop on
Water Infrastructure Sustainability and Adaptation to Climate Change

42

that they can focus on supervising engineers. From the perspective of a local government water resource manager, it is useful to have scientists defending the science so the policy makers do not have to. A climate change impacts researcher commented that several utilities and water districts have worked together as a "climate group" to bring together scientists and policy makers. These groups invite scientists to talk about various topics so as to inform the utilities on the science of climate change. Such efforts (e.g., NOAA's Regional Integrated Sciences and Assessments (RISAs)) are helping to fit all the relevant pieces together.

A hydrologist responding to the initial question of the importance of predictive tools commented on the importance of having real-time forecasting. The National Weather Service has been issuing real-time forecasting information from short-time intervals to seasonal intervals. In doing so, many researchers have demonstrated with clarity that if probabilistic rules as opposed to deterministic rules are adopted, there is a significant increase in efficiency.

A climate modeler asked how many water resource districts are completely unexposed to climate modeling/climate change. A local water utility manager responded that King County developed a guidebook to help local governments in their region to get started on these sorts of issues. Five years ago, there were not many local governments who were exposed to these ideas, but the tangible impacts of climate change have heightened awareness. Another water manager commented that large utilities have many more resources at hand, and the smaller ones are definitely less advanced. For many utilities, climate change adaptation is a new concept and there is much skepticism, but there is a change toward a direction of greater acceptance, particularly amongst smaller utilities.

A climate impacts researcher pointed out that when looking at national associations and trade groups, there are many that are on board with climate change initiatives, and there has been a lot of activity in the last few years. Many utilities now belong to at least one of these groups, if not more. A watershed manager responded that the extent of this sort of participation depends on the size of the utility and its location. Utilities in the Great Lakes region, for example, might have different reasons for considering climate change impacts than utilities on the East or West Coasts.

A water consultant commented that many small utilities use probabilistic operating rules. For example, Rocky Mount, North Carolina, does assessments every month as to the probabilities of its reservoir falling short. The State of North Carolina has purchased access to many tools that the public can use, and there is evidence of the public using them. Using these tools, local governments have access to more recent and up-to-date information. If they have the tools, they have the ability to move the information out into the real world of operations. In addition, if your operations tools are also your planning tools, you have the ability to use similarly formatted ensembled climate forecasts. But if you want to move in that direction, you should focus on operations first.

A watershed manager commented that it is often difficult to come up with probabilities for stream flow and precipitation. When looking at these probabilities and looking at water impacts, what does this sort of information do to address stream and water treatment needs? A hydrologist responded that applying climate change projections to long-term plans for water quality is not difficult to do for many parameters, and that it is possible to link to other water quality models (e.g., the generalized water loading function). A researcher responded that NOAA is currently writing documents in support of efforts to produce real-time water quality forecasts, but it is on the back end of the project timeline.

Another water manager suggested looking at how utilities responded to concerns about water security. He said the response was not effective and should be studied.

A water manager commented that one thing happening with local governments is that people are already addressing climate change issues through much more efficient use of water resources as a result of drought pressures. Utilities that have already been reducing demand and supply in areas affected by shortages might be putting themselves into more vulnerable positions, since they might not be as able to adapt to future changes. One member of the water management community commented that this is the first he has heard of the significance of water conservation during these discussions. He suggested that the community could use the savings from conservation to develop supplies to meet new growth in demand. He questioned the logic behind the assertion that utilities are making themselves more vulnerable by implementing conservation programs to reduce demand. A hydrologist commented on the fact that if a water utility manages its reservoir to always meet demand, it would run out of water. Reliability is a function of conservation. Rocky Mount, North Carolina, worked their way out of a shortage, not by building new pipelines, but by coming to an operating policy that looked ahead, effectively saying that when the risks got too high they would reduce demand from the reservoir. The questions to ask are, "What is reliable? How do you make sure that you are never going to run out of water?" Conservation makes it harder to do short-term measures if you have no slack in the system anymore. You had better have a larger reserve at the end of time, in case of devastating drought.

A water resources manager asked about joint climate and water community initiatives, and whether these initiatives have been focusing exclusively on water supply impacts from climate change and not looking at stormwater issues as well. A climate change impacts researcher responded that in his experience these initiatives are looking mostly at how climate change impacts affect supply. The reason why the focus has been on supply is because that topic is of primary concern to the utilities. Storm surges and flooding are often of secondary concern.

Proceedings of the First National Expert and Stakeholder Workshop on
Water Infrastructure Sustainability and Adaptation to Climate Change

44

6. Adaptive Management and Engineering: Information and Tools

This track focuses on the adaptive techniques, practices, and approaches used by water resource managers and engineers for dealing with current and potential climate change impacts. It addresses how water utilities look forward on a 10, 20, 50 year and longer planning window. The sessions focused on what climate change information and decision-making tools are used and will be needed to make management and engineering decisions in light of climate change.

6.1 National Infrastructure Condition Assessment and Adaptability

To begin the discussion on adaptive management and engineering, this session provided information on the current status of national infrastructure and some of the options for making decisions. The infrastructure built today will be in place for decades to come. The session addressed how climate considerations are included in investment decisions. Most water resource managers agree that there is now a level of actionable science on climate change and they are ready to move forward. However, decisions must be made in a framework of uncertainty. This uncertainty can be reduced, described, and even quantified. This session presents information on the available tools and techniques to assess infrastructure vulnerability and adaptability in the context of climate change.

Click here to read the remarks of the moderator (Dr Neil Stiber, EPA Office of the Science Advisor).

Rehabilitation, Replacement, and Redesign of the Nation's Water and Wastewater Infrastructure as a Valuable Adaptation Opportunity
Dan Murray, P.E., EPA National Risk Management Research Laboratory

Dan Murray began by stating that the 2002 Gap Report identified a $500 billion investment gap in infrastructure. In response, EPA developed the Sustainable Water Infrastructure Initiative in 2005. The initiative has four pillars, which include better management, water efficiency, full-cost pricing, and a watershed approach. It has an annual budget of $7 million and has several cross-cutting themes, such as innovation, partnerships, technology, and research. To understand the critical factors behind future demands and threats related to climate change, ORD initiated the Aging Water Infrastructure Research Program in 2007. The program is designed to facilitate the more cost-effective operation, maintenance, repair, and replacement of aging and failing drinking water and wastewater systems. It also facilitates the development and application of advanced designs and management approaches for drinking water and wastewater systems.

Mr. Murray identified many of the program's focuses. Primarily, the program focuses on understanding the critical factors behind future demands on and threats to our national water infrastructure systems. Other focuses include optimizing repair, rehabilitation, and replacement of drinking water and wastewater infrastructure, and extending the service life of in-place drinking water and wastewater system components. The program is evaluating the performance and cost of innovative technologies and approaches and investigating advanced system design and management concepts. It focuses on detecting, locating, and characterizing leaks in drinking water and wastewater conveyance systems, and designing systems with green infrastructure and low-impact development components to attenuate wastewater and stormwater flows.

Proceedings of the First National Expert and Stakeholder Workshop on Water Infrastructure Sustainability and Adaptation to Climate Change

45

The program's core focus is on the support of strategic asset management (SAM), which is a tool that can be used by water and wastewater utilities to adapt to the effects of global change. Through effective condition assessments of infrastructure systems and optimal investments in system rehabilitation, replacement, and redesign, especially focused on long-term system demands and threats, asset management can support successful adaptation to climate change. While strategic asset management is a tool for effective infrastructure adaptation, technical, and institutional challenges must be overcome to realize its potential. System modifications and resultant performance improvements will be measured over decades, making adaptation seem more like evolution.

The Aging Water Infrastructure Research Program's SAM uses a three-legged stool of tools: (1) condition assessment, (2) rehabilitation, and (3) advanced concepts. The condition assessment is key to determine how systems react, deteriorate, or fail. There are several key questions that must be answered during an evaluation of strategic assets. They include:

- What is the current condition of my assets (pipes, pumps, tanks, etc.)?

- What is my required level of service and current performance?

- Which assets are critical to sustained performance?

- What are the best options for investment in operations and maintenance (O&M), rehabilitation, and/or replacement to sustain long-term performance under climate change?

- What are the best, long-term, sustainable funding strategies?

Further questions must be answered while performing a condition assessment:

- What and where are my assets, and what is their current condition?

- What drives or will drive these assets to deteriorate and fail over the short- and long-term?

- What are my customer service demands, regulatory requirements, and current performance?

- What are the remaining useful lives of these assets and the performance consequences of their deterioration and failure?

Mr. Murray provided an assessment example performed by Melbourne (Australia) Water. Melbourne's assessment found increased potential for pipe corrosion in the wastewater collection system as a result of increased sewage concentrations associated with water conservation, increasing ambient and seasonal temperatures, and longer travel times within the system. They also saw increased incidence of sewer overflows due to increased rainfall intensity during storms, increased risk of pipe failure and collapse due to dry soil conditions, and finally increased salinity levels in recycled wastewater due to rising seawater infiltrating into the collection system and flowing to wastewater treatment plants.

Mr. Murray identified another tool that is vital for SAM, known as rehabilitation/replacement/ redesign. To evaluate the potential of this tool for current assets, managers must answer the following questions:

- What are the O&M, rehabilitation, replacement, and redesign options available to sustain performance?

- Which options are feasible given the condition assessment?

- What are the lifecycle costs of these feasible options?

Proceedings of the First National Expert and Stakeholder Workshop on
Water Infrastructure Sustainability and Adaptation to Climate Change

46

- What are the optimal choices that balance available resources with an acceptable risk of performance failure?

In conclusion, Mr. Murray states that asset management is a tested strategic framework for addressing water infrastructure adaptation challenges, but infrastructure adaptation will be more like an evolution as success will be measured across decades. Rehabilitation, replacement, and redesign decisions made today will affect system performance for 30, 40, 50, or even 100 years to come. There are many technical and institutional issues that must be addressed now to accelerate infrastructure adaptation, including the development of tools that enable local infrastructure adaptation decision making. In addition, accepted modeling and engineering practices need to be challenged, and innovative approaches and designs need to be tested.

Click here to view Mr. Murray's presentation.

Click here to read the transcript of Mr. Murray's remarks.

Flood Control and Surface Water Management Infrastructure in the Age of Climate Change
Dr. Rolf Olsen, U.S. Army Corps of Engineers Institute for Water Resources

Rolf Olsen began by asserting that there is a tradeoff between flood control and water supply and climate change is already affecting these tradeoffs. Water storage must allow for minimum flows, which can protect water quality, ecosystems, and passage levels. Much of the U.S. Army Corps of Engineers' reservoirs are used for recreation, which often develops into the largest concern for maintaining appropriate water levels. Warming has already driven observable hydroclimatic change, such as less snowpack and earlier snowmelt runoff. For example, a study of average stream flows in the North Fork River in California has found a significant earlier peak of discharge during the 1990s. These changes, coupled with increasing population in urban areas, have the potential to impact the design and operation of future drinking water treatment plants.

Flood storage space is funded by the federal government, so the U.S. Army Corps of Engineers does not own the water, just the storage space. Literature suggests that a warmer regime may result in about the same annual precipitation, but could result in less snowpack, earlier melt, flow shift, and greater storm variability/intensity. Thus the Corps must plan for more rain-flood space, particularly during the winter and allow an earlier reservoir fill. The Corps analyzed 22 GCM simulations for 2030 using projected temperature and precipitation ranges under two climate scenarios. It determined the 10th, 50th, and 90th temperature and precipitation percentiles by sub-basin and generated reservoir inflows using perturbed temperature and precipitation input using the National Weather Service River Forecast Center model. Finally, it tested flood control curves using Corps' ResSim model. The model shows that for New Bullards Bar in Yuba County, California, reservoir pool elevations exceeded flood pool zone during 19 of 144 samples and overtopped the dam during eight sampled flood events, indicating a need for more flood control space in the reservoir. A risk-based approach may be a better alternative for managers than the traditional rule curve.

Dr. Olsen discussed several adaptation strategies for water resource managers. The Intergovernmental Panel on Climate Change (IPCC) states that integrated water resources management (IWRM) should be the "instrument to explore adaptation measures to climate change." Adaptations to climate change include making better use of existing water resources. In order to

Proceedings of the First National Expert and Stakeholder Workshop on
Water Infrastructure Sustainability and Adaptation to Climate Change

47

accomplish this, Dr. Olsen described several strategies. Managers should evaluate if there are benefits from revising reservoir storage rules and authorized purposes as climate changes. They should identify new uses for reservoir storage that have a greater economic or social value than current uses, which may involve reallocation studies. Flood storage space could be evaluated based on updated hydrologic records and future projections. Storage space dedicated to maintaining ecosystems may become a priority for managers. An adaptive management process can have flexibility to adapt to observed climate conditions on an annual basis. Some projects may be operated more efficiently as part of a system of reservoirs rather than as a single project.

Two workshops were held in spring of 2007 that brought together California water managers to discuss climate change and reservoir operations. They attempted to answer the following questions: "What does climate change mean to California reservoir rule curves? When should a water control manual be modified?" How much modification can be done depends on the original Congressional authorization. The conclusions gathered from the meeting included a long-term goal to begin a dynamic, transparent process for updating rule curves, which should be a priority. However, Dr. Olsen noted that this is an expensive process. The managers also displayed a desire to increase flexibility in operations to improve system adaptability under climate change. A systems perspective should be employed that considers all objectives and integrates all operations. The current knowledge on climate change and variability may not be specific enough to adequately evaluate flood rule curves.

There are several reasons for managers to reallocate and re-operate their water management systems. Many existing water resources projects were designed decades ago and often used a relatively short hydrologic record that assumed stationarity. Current and future hydrologic conditions may also be changing for many reasons, including climate change, variability, and land use changes. Demographic, social, and ecosystem changes may result in changing uses for reservoir storage. Dr. Olsen concluded by stating that water control plans should be reviewed and adjusted, when possible, to meet changing local conditions. Changes in reservoir operations can be time-consuming and expensive, often requiring an environmental impact statement with public participation by stakeholders with different objectives.

Click here to view Dr. Olsen's presentation.

Click here to read the transcript of Dr. Olsen's remarks.

Climate Change Readiness Assessment and Planning for the Nation's Drinking Water and Wastewater Utilities
Dr. Stephen Buchberger, P.E., NRMRL-UC WRAP Team

Stephen Buchberger began by stating that water utilities with dwindling and compromised supply sources currently rely on outdated infrastructure. The aging infrastructure attempts to produce high quality drinking water, treated to meet increasingly stringent regulatory standards and to deliver finished water to a growing customer base of informed consumers with high expectations but few financial resources. Dr. Buchberger identified several issues that he explained are currently converging crises that could evolve into a perfect storm. These issues include a shifting climate, evolving institutions, a growing population, changing regulatory issues, uncertain economics, and aging infrastructure. Water utilities are seeing this perfect storm as a convergence of independent events that form an environment never experienced before.

Proceedings of the First National Expert and Stakeholder Workshop on
Water Infrastructure Sustainability and Adaptation to Climate Change

48

ORD and the University of Cincinnati developed the Water Resources Adaptation Program (WRAP) to perform a comprehensive survey to identify and analyze the most important factors affecting the performance of the nation's water resources infrastructure over the next 50 years. Dr. Buchberger identified several surveys were that already in existence, for instance, the EPA NEEDS Surveys, the U.S. Conference of Mayors City Water Survey, the WRF Strategic Assessment, and the American Society of Civil Engineers (ASCE) Infrastructure Report Card. The EPA NEEDS survey is a detailed estimate of the cost of construction of all needed publicly-owned treatment works in all of the states (where a *need* is a project, with associated costs, that addresses a water quality or public health problem). The 2005 U.S. Mayors survey found that aging water infrastructure is the number one "every-day" chronic issue, and a 2007 survey found that asset management programs are increasingly vital for water utility operation. The WRF Assessment found that trends in 2005 continue to be primary drivers of water utility strategies today, including aging infrastructure and financial constraints. It identified trends that are comparatively new or more pronounced since the 2000 assessment, including security and climate change.

The Five Cities Plus Regional Survey and the WRAP National Questionnaire were developed by ORD and the University of Cincinnati. The Five Cities Plus survey included Cincinnati, OH; Columbus, OH; Fort Wright, KY; Indianapolis, IN; Louisville, KY; and St. Louis, MO. It was completed by the directors of these municipal wastewater agencies in June 2008. The survey is similar to the U.S. Conference of Mayors Survey. The WRAP National Questionnaire was a 40-question web-based survey that was distributed on-line via several national water organizations, including the Association of Metropolitan Water Agencies (AMWA), the National Association of Clean Water Agencies (NACWA), and the National Association of Water Companies (NAWC) to a select subset of the nation's drinking water and wastewater utilities during the summer of 2008. The main objective of the questionnaire was to identify, through the eyes of the water industry, the most important factors likely to affect the performance and sustainability of the public and private water resources infrastructure across the United States over the next 50 years. A total of 55 water utilities responded (31 drinking water agencies and 24 wastewater agencies), representing nearly 43 million customers with infrastructure assets that included 91 water treatment plants, over 520 storage tanks, nearly 1,200 pumping stations, and over 73,000 miles of pipeline. Four out of five respondents expected demand for water service to increase over the next 20 years with an average annual growth rate of about one percent. The survey results showed that most agencies had developed a formal master plan and planning horizons ranged from 5 to 40 years with a median of about 20 years. However, nearly 40 percent of water utilities did not have a formal asset management program. Of the four primary water-related infrastructure categories (i.e., pipes, pumps, tanks, plants), the pipeline systems used to distribute drinking water and collect wastewater were judged to be in the worst condition. The generally poor self-assessment of existing urban water piping systems is consistent with the overall low grade assigned to drinking water and wastewater in the recent ASCE report cards on the nation's infrastructure. The survey found consistencies among the wastewater and drinking water utilities and the self assessment portion of the survey shows that both groups feel they are performing better than the ASCE Report Card shows.

Dr. Buchberger mentioned the top three challenges having the greatest impact on the operation and performance of the nation's water industry over the next 50 years: (1) aging infrastructure, (2) government regulations, and (3) funding shortfalls. This ranking was consistent for agencies in the drinking water group and in the wastewater group. While climate change was recognized as an impending issue, it was viewed as a distant concern in comparison to the more immediate and urgent operational needs of the water utility. The industry-wide practice of developing and updating a master plan provides a ready opportunity for incorporating flexible mitigation and adaptation strategies to help water utilities cope with anticipated impacts from global climate changes. Dr. Buchberger concluded that the WRAP survey shows that aging infrastructure and climate change are complementary concurrent challenges for the water industry. There is a tremendous opportunity to rethink and redesign the water infrastructure, reflecting adaptation to climate change, low impact development (LID), decentralized systems, green options, and sustainable approaches.

Click here to view Dr. Buchberger's presentation.

Click here to read the transcript of Dr. Buchberger's remarks.

Assessing the Impacts of Climate Change on Drinking Water Treatment
Dr. Robert Clark, P.E., NRMRL-UC WRAP Team

Climate change may affect both surface water and groundwater quality. Increases (or decreases) in precipitation and related changes in flow can result in problematic turbidity levels; increased levels of organic matter; high levels of bacteria, viruses, and parasites; and increased levels of pesticides in lakes, rivers, and streams. Some areas may experience droughts resulting in elevated levels of potentially toxic algae, high concentrations of organic matter, and bacteria. Climate change coupled with population changes may therefore impact existing and future drinking water treatment infrastructure. Some of these impacts have the potential for causing serious violations of drinking water standards. These changes, coupled with increasing population in urban areas, have the potential to impact the design and operation of future drinking water treatment plants. Robert Clark states that drinking water treatment has the following three general objectives: (1) to remove any toxic or health-hazardous materials, (2) to remove or inactivate any disease-producing organisms, and (3) improve the aesthetic acceptability of the water. Each goal must be achieved at a reasonable cost.

The EPA Water Treatment Plant (WTP) Model has been adapted and utilized to address these impacts. It has been developed to assist utilities in identifying and screening new treatment technologies for meeting new and existing regulations. It also assists utilities in evaluating the possible effects of source water or treatment process operations on disinfection by-product (DBP) formation. The WTP Model uses empirical correlations to predict central tendencies of variables such as Natural Organic Matter (NOM) removal, disinfection, and DBP formation in a treatment plant. The model has been validated using data from the EPA's Information Collection Request (ICR) in conjunction with data from the Greater Cincinnati Water Works' Miller (surface water) and Bolton (ground water) treatment plants. It predicts changes in water quality parameters caused by chemical addition and/or by treatment configuration specified by the WTP Model. The team has modified the WTP Model to accept real-time inputs.

To illustrate the model's application, an example has been constructed using historical total organic carbon (TOC) data and two hypothesized increases in TOC (to represent climate change impacts). The effects of these three scenarios were evaluated by simulating a conventional water treatment plant and a conventional plant with the addition of granular activated carbon (GAC). The model assesses the regulatory impact on total trihalomethanes (TTHMs) through conventional treatment and examines possible treatment modifications to achieve regulatory targets. It was found that the conventional plant could not meet current drinking water standards under these scenarios. However, by adding GAC and varying reactivation frequency, it was found that the simulated GAC plant could meet current DBP regulations, but at increased cost.

In conclusion, Dr. Clark states that climate change will most likely have an impact on surface water quality. The WTP Model allows an evaluation of the impact of water quality changes on water

Proceedings of the First National Expert and Stakeholder Workshop on
Water Infrastructure Sustainability and Adaptation to Climate Change

50

treatment. The model can show quality impacts in real time and simulate effects of quality and population growth. Finally, it can estimate cost impacts of design and operational changes.

Click here to view Dr. Clark's presentation.

Click here to read the transcript of Dr. Clark's remarks.

Summary of Discussion Session

A participant stated that large utilities have been discussed a lot, but we are not applying these concepts to small utilities. Gloucester, MA, has the highest rates in the country and is not thinking about climate change. There is a need to engage the small- to medium-sized utilities.

A water manager mentioned that there are challenges with water treatment and many different strategies, particularly surface water versus algae. Does the model address different issues? A water researcher answered that the model makes some adjustments for different issues, but it still needs some enhancement.

A water resources consultant mentioned that modeling for the Southeast, including that done in South Carolina, indicates that operating rules need to be changed. They asked whether this is this more of a political issue. A water manager responded that there are different priorities in the Southeast for the Corps. The questioner then asked if climate will be part of the discussion and the water manager stated that he hopes so.

A scientist stated that there is a confluence of aging infrastructure and climate change which is leading to a golden opportunity. Is it possible to present them to taxpayers as one issue? Steve Buchberger answered that it is a question of economics; plans for rehabilitation need financial plans. It could be adequate to simply provide funding for green infrastructure. The moderator stated that perhaps cost is an issue, but uncertainty is a key issue.

A water manager provided an example of Jasper Water and Sewer. When utilities do renewal and rehabilitation work, it is done piecemeal because of financial restraints. Ten miles of pipe are not replaced, just sections. Components of treatment plants are replaced, not entire plants. Utilities are constrained by these financial issues.

An urban planner stated that Keene, NH, produced a climate action plan for adaptation in November 2007. They deal with water infrastructure through capital improvement programs. Their approach was to first work on mitigation issues that also addresses adaptation. They have discovered a lack of decision making and financial tools available to local governments. She also mentioned that conversations on adaptation are difficult to have with communities. A water researcher responded that this brings up the issue of how communities pay. It is important to explain to them the expected damages if they do not pay.

A member of the water research community commented that when we look at what we are investing now versus what is needed over the next 20 years, there is a $540 billion difference. We could reduce this cost by up to 20 percent if everyone implements best practices. People will end up paying two times as much in the future as they do now, and that is a low-ball figure. Having a conversation with people is vital. You must tell them that maintaining water infrastructure is costly. There are three things utilities should demonstrate: (1) people must understand what you do, (2) they have to see the value, and (3) you must demonstrate that you are using best practices. Some

Proceedings of the First National Expert and Stakeholder Workshop on
Water Infrastructure Sustainability and Adaptation to Climate Change

51

sophisticated utilities are using asset management strategies, but there are many more utilities who are not close to this management practice and many who do not even know the current state of their assets. Another member of the water research community mentioned that there is a lot more involved than just engineering. We will be adding over 100 million people to the United States by 2040. Where will all these people go? All other practices must be involved in this discussion, including environmental, economic, and health fields. This is not a free resource.

Another water researcher stated that mitigation needs money in the form of a price on carbon. How does adaptation get involved in the cap-and-trade discussion? A member of the water management community stated that there is a lack of leadership with this issue. We have an entitlement populous; everyone does not have accountability for their actions. Engineers must redesign, and politicians must revalue because the public must be given many different options.

6.2 Progressive Adaptation: Planning and Engineering for Sustainability

Water utility management inherently is about making decisions under uncertainty for ranges of outcomes. Water managers have had to cope with change and confront risks. Water managers routinely look at the probability and the likelihood of an event; they look at the consequences or impacts of an event; and finally, they look at risk mitigation or risk reduction or avoidance. This session covered some tools and approaches that have been used by water managers to address risks and some that are being developed to help water managers also address risks from climate change.

Click here to read the remarks of the moderator (Steve Allbee, EPA Office of Wastewater Management).

Overview of Integrating Climate Adaptation into Lifecycle Costing and Planning
Steve Allbee, EPA Office of Wastewater Management

Steve Allbee began by stating that in order to integrate climate adaptation into lifecycle costing, managers need to address the following questions to determine critical facilities: "How are they affected? What is the likelihood of the effect (although they cannot determine when or how large the affect will be)? What does it cost to mitigate risk? What are the consequences of not mitigating risk?" Identifying critical facilities requires a risk-driven assessment that asks what the probability is and what is the consequence. Managers should focus management and resources toward high probability and high consequences. Adaptation requires risk exposure management applied to the probability or likelihood of event, the consequence or impact of event, and risk reduction and avoidance (also known as risk mitigation).

Mr. Allbee describes a stepped approach to conducting assessments. A 100 percent level (basic) requires at a minimum a glance at all assets, while a 20 percent level (intermediate) focuses on some assets that need more thought. Finally, a 5 percent level (advance critical) requires a full economic analysis, which is very costly. Mr. Allbee identified two key questions: (1) Is the impact reasonably predictable? (2) Is it cost effectively preventable? Resulting management strategies could be very different. In order to improve the confidence level, the best appropriate process and quality of data used will result in confidence that the course is the right one. However, most utilities are not willing to invest the money into such an analysis. A strategic level map of organizational risk should not display any assets in a critical risk area. When evaluating cost perspectives, managers must look at direct lifecycle costs and economic costs. Direct lifecycle costs include acquisition,

Proceedings of the First National Expert and Stakeholder Workshop on
Water Infrastructure Sustainability and Adaptation to Climate Change

52

operation, maintenance, renewal, disposal, and decommissioning. Economic costs include financial costs, direct costs to the governmental organization, direct consumer costs, and finally, an analysis of the triple bottom line. Mr. Allbee concludes that after integrating initiatives into the management framework through the asset management business processes, managers can commit resources to actual spending from the operational and capital budgets.

Click here to view Mr. Allbee's presentation.

Click here to read the transcript of Mr. Allbee's remarks.

Adaptation of Water Infrastructure Investments to Changing Demands and Climate Variability: A Systems Approach

Dr. Vahid Alavian, World Bank

Vahid Alavian began by stating that water is an integral element of many economies and water investments influence the economy at both the macro and the micro levels. The impacts of climate change on the hydrologic cycle are felt in similar ways by both utilities and developing countries. The degree to which a water system is susceptible to, or unable to cope with, adverse effects of climate change defines its vulnerability to climate variability and extremes. A number of factors make water investments in many developing countries vulnerable to the impacts of climate change and may as a consequence expose the country to unmanageable economic shocks.

At the cross-cutting level, climatic impacts will have significant consequences on infrastructure systems that deliver services and/or manage the resources. These systems include water, energy, transport, and ecosystem management infrastructure. Infrastructure system here is defined as an integrated system of physical, institutional, and financial elements. Regardless of the purpose, these systems are intimately linked to water. The design of these systems will have to change for both developed and developing countries. At the sector level, water systems can be classified as those that deliver water services (e.g., water supply and sanitation, urban drainage, wastewater, irrigation) and those that help manage resources (e.g., multi-purpose infrastructure, watershed management, river basin management). Water services and water resources are both affected very differently by climate change. These systems are already under pressure as a result of increasing water demand through rapid urbanization, degrading infrastructure due to lack of maintenance, and weak management institutions in many countries. Climate change is expected to seriously jeopardize the water systems and it is vital for developing countries to transition into a more manageable system.

Estimates show that by 2025, 2.8 billion people, and by 2050, 3.4 billion people, will be living in primarily water stressed watersheds globally, as compared to the current 1.4 billion. The rapid growth of slums is also a major area of concern because high density and inadequate urban planning make the provision of sanitation services in slums a particularly difficult challenge. About 75 percent of the population growth over the next 15 years will be in cities of less than 5 million inhabitants, with over 50 percent in cities fewer than one million where services are already in short supply and of poor quality. Many of these large cities are coastal cities which face additional climate change impacts.

The key challenge is financing investment in water systems. Annual investment required to meet the millennium development goals is $25–30 billion and current levels are only at $15 billion. Public spending on infrastructure was halved between the early 1980s and the late 1990s. Currently, it is

Proceedings of the First National Expert and Stakeholder Workshop on
Water Infrastructure Sustainability and Adaptation to Climate Change

53

at about 2 percent of gross domestic product (GDP). Investment in water supply and sanitation (WSS) is less than 1 percent. Expectations that private investment would compensate the fall in public funding have not been met. The current investment level is $1 to $2 billion per year, primarily focused on specific markets (e.g., China, treatment facilities). Overseas development assistance to the water sector has declined since the mid-1990s; the share of aid allocated to WSS is currently 6 percent. The current financial crisis has cut deep into aid and investment in WSS and other infrastructure projects.

There is already a lot of hydrologic variability in the developing world and climate is not as large a concern as consistent water supplies and sanitation. A winter flood in Kenya in 1997–1998 caused $2.39 billion in infrastructure damages, while a drought between 1998 and 2000 caused $2.4 billion in losses. They had a 22 percent impact on GDP per year, in contrast to Hurricane Katrina, which had losses of less than 0.5 percent of GDP. Infrastructure service delivery is at the core of water security, energy security, and climate change agenda. Adaptation is critical for irrigation, water supply, and hydropower.

Dr. Alavian stated that "no regrets," "good practice," and "sustainable" actions can be justified with or without climate change. These actions include demand management, efficiency, productivity, and intelligent and flexible design and operation of water infrastructure (including "on demand" intervention, and infrastructure that "scales to needs"). "Climate justified" actions are different because they require more care in measuring impacts, more precise assessment of system vulnerability, a deliberate decision on the degree of risk to be taken, and strong justification of usually high additional costs. "No regrets" adaptation measures for infrastructure include intelligent and flexible design and operations. A major effort is needed on rehabilitation, including cross-sectoral infrastructure (e.g., water, energy, and transportation), early warning mechanisms, and an increased capacity to respond. Improvements in monitoring and assessment technology, efficiency improvement, and demand management are also key. Dr. Alavian concluded that evaluating economics and tradeoffs include decision making under increased uncertainty and risk-based project economic analysis. Financing mechanisms include risk insurance (for systems and for customers, notably the poor) and incentives for private sector investments.

Click here to view Dr. Alavian's presentation.

Click here to read the transcript of Dr. Alavian's remarks.

A Review of Quantitative Methods for Evaluating Impacts of Climate Change on Urban Water Infrastructure
Dr. Walter Grayman, P.E., NRMRL-UC WRAP Team

It is widely accepted that global climate change will impact regional and local climates and alter some aspects of the hydrologic cycle, which in turn can affect the performance of the urban water supply, and wastewater and stormwater infrastructure. How the urban water infrastructure will be affected and how these impacts may be mitigated by design or operational changes has been the subject of study and much conjecture.

In a qualitative assessment, potential impact pathways are identified and a general assessment of the relative significance of the pathways is made. These pathways must be understood before one can move to quantitative assessments. A quantitative assessment extends this analysis to include mathematical modeling techniques to calculate numerical estimates of the impacts of global climate

Proceedings of the First National Expert and Stakeholder Workshop on
Water Infrastructure Sustainability and Adaptation to Climate Change

54

change on the hydrologic cycle, water quality, the ecosystem, land use/population development, and ultimately on the actual performance of the urban water infrastructure. Direct impacts include decreases in precipitation, while indirect impacts include the deterioration of water quality via temperature changes; affects on insect populations; and impacts of insects on trees, leading to soil erosion. Other issues include uncertainty, timeframe, population growth, and energy stressors.

Dr. Grayman asked, "How does climate change affects infrastructure interaction?" One must model ecosystem changes, hydrologic and water quality processes, and land-use and population changes. Various metrics, such as cost, flow quantities, flow quality, population impacted, etc., can be used as measures in a quantitative analysis. There are many direct and indirect impact pathways and feedback loops associated with these processes. An important component of this type of analysis is an assessment of the uncertainty associated with the quantitative estimates. Developing a framework for evaluating the quantitative impacts of climate change on the urban water infrastructure using existing modeling techniques representing the processes and interactions is the ultimate goal. Model categories include climate change, hydrologic models, water quality models, ecosystem models, population/land use models, infrastructure models, and systems dynamics models (which are integrations of all models).

Climate change models (GCMs) are mathematical models of the Earth's climate. Coupled climate models (AOGCMs) represent the interactions between the atmosphere, ocean, land surface, and sea ice. Spatial or temporal downscaling derives regional climate data from coarse-resolution model outputs. Some of these issues surrounding climate models are uncertainty and the inability to account for short-term intense precipitation events. Hydrologic models are rainfall-runoff models that predict streamflow resulting from precipitation. They can either be in time scale (event based, such as a particular storm, or continuous), or spatial scale which analyzes large or small watersheds. Hydrologic models can be integrated with geographic information systems (GIS), however, most pre-date climate change applications. Water quality models measure how water quality varies temporally and spatially due to loadings and the environment. They are often used in conjunction with hydrologic models. Examples include stream water quality models (QUAL2K, WASP), integrated hydrologic/water quality models (BASINS, SWMM), integrated streamflow/water quality models (EPD-RIV1), and integrated groundwater flow/water quality models (MODFLOW).

Climatic conditions determine where individual species of plants and animals can flourish, therefore ecosystem models can be used to simulate changes in processes and geographic distributions. Land use/population models project temporal and spatial changes in land use and population. They are often integrated with GIS and attempt to measure feedback from climate change on future land use and population patterns. Infrastructure models predict the performance of components of the urban water infrastructure. Examples include water distribution systems (EPANET), urban stormwater systems (SWMM), water treatment models (WTP), and wastewater treatment models. All of these models have issues surrounding uncertainty, reliability, and interactions with climate change.

Systems dynamics models test "what if" scenarios of complex systems with feedback loops. Examples include the MIT Greenhouse Gas Emissions Simulator, the C-ROADS: The Climate Rapid Overview and Decision-Support Simulator, and the ASU Systems Dynamics analysis of urban vulnerability to climate change. A question surrounding systems dynamics models is whether they can describe the processes in sufficient detail and accuracy to trust the results of the models. In order to help solve these issues with the above mentioned assessment framework, Dr. Grayman identified several research needs, that there should be an enumeration of qualitative pathways between climate change and infrastructure impacts, and there should be an in-depth evaluation of quantitative assessment tools of the above mentioned models.

Dr. Grayman concludes by stating that we must have a clear understanding of the mechanisms and pathways by which climate change can impact the performance of the urban water infrastructure. Existing quantitative models must be evaluated to ensure that they adequately represent these mechanisms and pathways. If not, further research and development is needed. Finally, he stated that there will always be a large degree of uncertainty in the modeling process associated with climate change.

Click here to view Dr. Grayman's presentation.

Click here to read the transcript of Dr. Grayman's remarks.

Water Use and Re-Use in Energy Technologies in a Carbon-Constrained World
Dr. Pratim Biswas, P.E., Washington University in St. Louis

Water infrastructure sustainability and adaptation to climate change is a very broad topic. Clean and alternate energy production technologies are being developed and implemented. A key contributor to climate change is fossil fuel based energy production, which means we are now in a "carbon-constrained world." Water issues related to clean energy need much more attention than they are currently receiving. There are several energy scenarios that need to be discussed. Energy sources form a "mixed bag" including fossil fuels and alternative energy. Population growth is placing stress on energy production, particularly in developing countries. Population growth and energy growth go hand-in-hand, so we need technological innovations to step up energy production in an environmentally benign manner.

Interests in energy usually come and go. Alternative energy gained importance in the 1970s, but that interest quickly waned. Pratim Biswas introduced the energy equation, which includes fuel economics and reserves, energy security, and carbon dioxide (CO_2) and other environmental emissions. There are many technological issues and challenges as we transition into a clean energy economy. Society will rely on fossil fuels in the interim (next 50 years). Intermediate term solutions include methanol and other fuel cells, hybrid energy generation, and biofuels. Long-term solutions are hydrogen and distributed energy production.

Water and energy are intimately interconnected. It is vital to carry out a "holistic analysis" which accounts for environmental factors such as water use. For example, during electricity production using coal, a holistic analysis examines the water use during mining, electricity production, and waste treatment. It will analyze their operational use of water, then use tools to see if water can be conserved or "re-used." Each 500-MW coal plant burns 200,000 kg of coal per hour, releases 750,000 kg of CO_2 per hour, and uses 2.2 billion gallons of water per year (equivalent to a city of 250,000 people). Fifty percent of electricity in the United States comes from coal, amounting to two billion metric tons of CO_2 emissions per year. The U.S. Department of Energy (DOE) has performed detailed studies on water use in coal plants.

Many of the alternate technologies will have varied uses of water, and this has to be considered as choices are made for the future. Solar energy is diffuse but plentiful, and has the potential of being used for distributed generation. While silicon has been a mainstay of the semiconductor industry, it is expensive and energy intensive to produce (current photovoltaics (PV) have attained efficiencies of 20 to 25 percent). Solar energy needs low-cost production methods and materials that are plentiful, such as oxide semiconductors. The water required to produce hydrogen for a U.S. fuel cell vehicle fleet is around 100 billion gallons of water per year. We currently use about 300 billion

Proceedings of the First National Expert and Stakeholder Workshop on
Water Infrastructure Sustainability and Adaptation to Climate Change

56

gallons of water per year in the gasoline refinery industry alone. Domestic water use in the United States is about 4,800 billion gallons per year. The United States uses about 70 trillion gallons of water per year for thermoelectric power generation. Fossil production of electricity consumes about 0.5 gallons of water per kWh produced. Wind and PV consume no water during their electricity production. This means that every kWh of wind that replaces a kWh of coal saves 0.5 gallons of water. If we aggressively install wind, then our overall water usage would drop. Bioenergy faces the issue of food versus energy crops. Dr. Biswas and Washington University have developed an auditing process for analyzing water used during the ethanol production process. This is available at: www.aerosols.wustl.edu/education/energy/ethanolaudit/. The greening of buildings, both existing and new infrastructure, is another energy goal. Washington University is trying some novel approaches in a new state-of-the-art laboratory building that will reduce water use and take advantage of water recycling and reuse.

In conclusion, Dr. Biswas states that holistic, integrated analysis is a key for understanding the relationship between energy production and water issues. We need to develop auditing tools to evaluate energy and water usage, and then use these tools to design water re-use and conservation to guide the overall design of an energy production system in a "carbon-constrained world." The issue of water will continue to be very important.

Click here to view Dr. Biswas' presentation.

Click here to read the transcript of Dr. Biswas' presentation.

Summary of Discussion Session

A member of the water research community mentioned that distributed energy offers potential solutions that have parallels to the water system. WEF members could be more interested in this. Another water researcher added that tools need to be developed to measure the economies of scale. These tools are being used in the energy industry and they can be applied to the water industry as well. A water manager answered that the big difference is that you can store one and not the other. A participant remarked that in the 1970s we went to a centralized system to deal with pollution, now it seems like the opposite. A member of the water management community added that Chicago is considering the terrorism threat for decentralized systems. We need to find multiple sources of water, not just Lake Michigan, in case our treatment system went down. A water resources manager mentioned that we need to take into account the dimensions of both centralized and decentralized systems.

A water manager stated that they are finding that the private sector is involved in reclaimed water used for irrigation purposes. Regulators are not prepared for those who are operating facilities in private developments.

Another water manager mentioned that they are also looking to diversify their sources, looking at desalinization on the Hudson River. Dealing with the public perception surrounding this is difficult. A water manager remarked that the five-year drought in Australia led to a brand new assessment of priorities, which in turn led to desalination.

A water researcher stated that he wanted to know more about the asset management map. A member of the water research community responded that the example came from Orange County, CA, in which the organization engaged on where risk was located and found that it did not fall at

Proceedings of the First National Expert and Stakeholder Workshop on
Water Infrastructure Sustainability and Adaptation to Climate Change

57

critical points. There should be an action plan for each risk. He stressed that this was very important.

A water manager asked, that while risk assessment was relied on by the energy sector, how do they protect against the risks of an energy system failure? Another member of the water management community added that Miami cannot depend on the power grid to operate due to hurricane risks. Miami water created redundant power supplies at 1,000 pump stations. They have an operating agreement with the utility and can use excess power. They have diesel-fired boilers, which are not ideal from an environmental standpoint. In regards to the relocation situation, Miami has assets at risk due to sea-level rise, so at what point do they discuss relocating the population because of climate change? A water manager responded that environmental refugees are a concern that the World Bank is looking at, but this runs counter to World Bank safeguards that all refugees must be moved to a better situation. A water resources manager remarked that we should copy Australia's water sector and focus on global collaboration on these issues.

A water resource consultant added that in terms of financing, these are tough decisions that ratepayers or regulators are not ready to make. Additional funding for research and development (R&D) and implementation of water infrastructure products is necessary.

A water manager responded by asking whose money is the issue here? Research at the federal level is dedicated to finding the most cost-effective resources and applying them across the country. The regulatory system has to evolve, and this is not an either or discussion between investment and research. One participant questioned the number of years people are willing to stick with a program. Research can lead to long-term, sustainable practices.

6.3   Adaptation Practices and Tools – Part I

The following two sessions discuss the adaptive management, engineering, information, and tools, available to water utility managers. A wide range of options is illustrated in Part I, including a focus on operational adaptations in California, a focus on potential institutional adaptations along the Ohio River; and a focus on comprehensive risk analysis as an action mechanism in the Boston area.

Click here to read the remarks of the moderator (Josh Foster, Center for Clean Air Policy).

Alternative Water Supply and Drinking Water System Operations: Preparation for Climate Change Adaptation in East Bay MUD
Dennis Diemer, East Bay MUD, CA

Dennis Diemer began by stating that climate change is a growing threat to communities that have enjoyed plentiful water supplies for decades and planned their water systems based on historical water supply records. Climate change science continues to evolve and provide a better understanding of potential impacts, which include lengthening periods of increased ambient temperature, rising sea levels, reduced snow blankets in the West, and rising river temperatures. Water managers face difficult decisions on how best to adaptively plan and reliably provide for the water needs of future generations.

The East Bay Municipal Utility District (EBMUD) has been closely following climate change science to understand the potential impact to the region. EBMUD serves 23 cities and over 650,000 customers. EBMUD has been using data and information on global and regional climate change impacts, and

*Proceedings of the First National Expert and Stakeholder Workshop on
Water Infrastructure Sustainability and Adaptation to Climate Change*

58

has applied them toward its water supply and operation planning efforts. EBMUD has cooperated with the Lawrence Livermore National Laboratory and modeled potential water temperature increases in the Mokelumne River and EBMUD Reservoirs. It has conducted hydraulic and hydrologic modeling to determine the potential impact of climate change on river flow and reservoir filling. EBMUD is tracking both the IPCC and State of California work to predict climate change impacts on the western United States. EBMUD's research identified six key areas of potential vulnerability: water supply, flood management, water demand, sea-level rise, power generation, and water quality.

There are a significant number of predicted climate change impacts on EBMUD's water supply, including varying hydrologic reliability. There will be more dry years, less snowpack, and earlier runoff. It is not known if there will be more or less precipitation overall, but less snowpack is very likely. For EBMUD, three out of ten years are dry years, and a 20 percent reduction in precipitation would increase the frequency of dry years to five out of ten years. There will be an increasing demand for water and increasing water temperatures. Customer use of water for outdoor irrigation and indoor water consumption may increase with a warming climate. Other effects include the lengthening of the growing season, and a decrease in soil moisture content. Warmer water temperature in river and reservoirs will have large effects on local fisheries and on water quality. The effects on water quality could lead to increased turbidity, decreased water treatment plant capacity, and increased water production costs. Infrastructure reliability will also be affected due to sea-level rise and delta vulnerability. EBMUD's power generation could also be affected according to the modeling. EBMUD could see increased energy demand, increased peak energy use of between 4 and 19 percent, lost power generation by 10 to 30percent, and lost revenues.

EBMUD has identified and developed appropriate adaptive strategies to prepare for the impacts of climate change. Climate change considerations have been accounted for in EBMUD's strategic planning process to support short-, intermediate-, and long-term decision making, including planning for its future water supply needs. Their planning approach has a climate change objective in 2008 which included a monitoring and response plan. The Water Supply Management Program (WSMP) 2040 incorporates the new Freeport Project with the Sacramento River supply. EBMUD developed water supply alternative portfolios that considered flexibility, diversity, reliability, carbon footprint, and energy consumption. EBMUD conducted sensitivity analysis to evaluate each alternative portfolio's ability to adapt to climate change The alternative portfolios included increased conservation, increased recycled water, water transfers, groundwater banking, desalination, and increasing the height of reservoirs. To increase GHG mitigation, EBMUD has conducted GHG inventories since 2005 and is certified by the California Climate Action Registry (CCAR). Onsite renewable energy such as solar PV and wind from microturbines have been installed. EBMUD has also begun a wastewater cogeneration project, in addition to its existing hydropower generation. Finally, a hybrid vehicle fleet has saved 12,000 gallons of gasoline and reduced CO_2 emissions by 103 metric tons.

Mr. Diemer concluded that EBMUD will continue to monitor the science and perform the necessary studies to evaluate the possible impacts of climate change. They have determined vulnerabilities and identified future actions to reduce them. EBMUD has incorporated the potential for climate change into EBMUD's strategic planning process and planned for water supply uncertainty with a diverse and flexible portfolio. Finally, they are implementing mitigation elements and taking a proactive approach to reducing GHG emissions.

Proceedings of the First National Expert and Stakeholder Workshop on
Water Infrastructure Sustainability and Adaptation to Climate Change

59

Click here to view Mr. Diemer's presentation.

Click here to read the transcript of Mr. Diemer's remarks.

Stormwater Management and Extreme Precipitation: Protecting Surface Water and Source Water Quality in Ohio River Watersheds

Alan Vicory, Ohio River Valley Water Sanitation Commission

The Interstate Compact of the Ohio River Valley established the Ohio River Valley Water Sanitation Commission (ORSANCO) to abate interstate water pollution in the Ohio Valley drainage. ORSANCO establishes discharge requirements on the Ohio River, and monitors and assesses water quality and biology. ORSANCO also respond to spills and coordinate the implementation of the Clean Water Act (CWA) and Safe Drinking Water Act (SDWA) Programs. The commission also conducts research and works closely with river users (particularly publicly owned treatment works (POTWs) and WWTPs). The Ohio River Basin is highly engineered and has 20 dams and 49 power generating facilities. Annually, it transports 230 million tons of cargo 981 miles from Pittsburgh, PA, to Cairo, IL, and is often called the U.S. industrial artery. The drainage basin covers 203,900 square miles in 14 states and is home to over 25 million people. It provides drinking water for 5 million people (29 intakes) and contains over 120 species of fish. The river is shared by six states and has a very diverse land use, including mining, agriculture, chemical industries, and coal production. Given the magnitude of the basin and the wide diversity of land use and hydrologic conditions, the challenge with respect to the study of predictive effects of climate change and formulation and delivery of mitigating and adaptive strategies are highly complex technically, politically, and institutionally.

From the standpoint of water quality, it is expected that climate change impacts on the Ohio River will be comparatively less severe than on most other streams, due both to its size and the extensive in-place infrastructure that facilitates the river's use for industrial navigation. Still, given the current general understanding that climate change will result in increased occurrences of drought and elevated flows, there are important implications with respect to water quality. For example, low flow conditions and higher stream temperatures heighten the risk of algae blooms and low dissolved oxygen, and higher observed levels of total dissolved solids, chlorides, and sulfate. In the event that more flow augmentation is necessary via U.S. Army Corps operated reservoirs, lower water levels may diminish recreational opportunities. Water quality effects from higher flows will likely include increased sedimentation, higher observed levels of polychlorinated biphenyls (PCBs) and dioxin (shown to increase with higher flows), increased bacterial levels due to sewer system overflows, and additional inputs of non-point related pollutants such as bacteria, atrazine, and nutrients due to agriculture production. Ten percent of the combined sewer overflows (CSOs) in the country are affected by additional bacteria loadings from high flows. Related high flow impacts include aquatic habitat impairment and hypoxia in the Gulf of Mexico. There are also important implications regarding spills. There are over 600 industrial discharges and chemical spills can show up 150 miles away from their release point. In high flows, spill events tend to be infrastructure-based, such as barge breakaways, while in low flows, spill impacts are exacerbated due to less dilution.

ORSANCO is new to the issue of climate change, but as the commission engages the issue, it expects that its role, activities, and relations with its member states and federal agencies will follow current policy. Mr. Vicory stated that the commission needs to know what to track with an established monitoring network and sound historical data. Is there sufficient plasticity in current regulatory processes to allow us to keep pace with pace of change? Should our criteria change as

Proceedings of the First National Expert and Stakeholder Workshop on
Water Infrastructure Sustainability and Adaptation to Climate Change

60

climate changes? If not, the alternative might be that we list more impairments. In February, ORSANCO is holding a scenario planning, or "futures visioning" workshop. The commission is in need of predictive tools for climate change and a clearinghouse for these tools. It will also be important to identify interstate management implications and integrate the interaction of energy and water into the climate management challenge. This will be a particularly key issue for the Ohio Valley.

Click here to view Mr. Vicory's presentation.

Click here to read the transcript of Mr. Vicory's presentation.

Case Study: Risk and Management Analysis for Progressive Adaptation of Water Supply in Metro Boston

Dr. Paul Kirshen, Tufts University

Paul Kirshen stated that proactive adaptation requires that actions are taken before major impacts have occurred. Research has shown that this is a more cost-effective approach than reactive adaptation. The requirements of proactive adaptation planning include a consideration of climate change in current planning so proper adaptation can occur, in particular, explicit accounting for climate non-stationarity and other uncertainties. There is a need to consider dynamics, i.e., the timing of adaptation actions in recognition of the impacts of other driving forces on infrastructure and the environment (e.g., population, globalization, urbanization); GHG mitigation; active stakeholder involvement at all levels; and consideration of adaptation interactions among various infrastructure and environmental sectors. A scenario-based risk assessment procedure addresses many of these challenges. The consideration of system performance over all conditions moves away from the traditional "design event" approach (i.e., 10-year storm, 100-year storm). It focuses upon "residual risks," i.e., consequences of impacts that occur when the design level of a project is exceeded and explicitly recognizes that uncertainty (lack of quantified probabilities) exists in the process, which must be addressed through scenario analysis. The assessment relies upon two-way communication with stakeholders to select the level of risk they can tolerate while considering tradeoffs of multidimensional costs versus safety.

Scenarios of outcomes that cannot be characterized by known probabilities can be combined with a risk-based approach that is used for actions whose outcomes can be characterized by a probability. In the risk-based approach, the costs and benefits of the performance of an infrastructure system over the entire range of possibilities are evaluated. The complete analysis can be assembled similarly to a decision tree. If probabilities can be assigned to the uncertainties, then the analysis collapses to a normal decision tree. The major uncertainties in adaptation planning may include various land use, GHG emission scenarios, the climate impacts of each GHG scenario, and local socioeconomic scenarios. The outcomes of scenario-based risk assessment analysis are probabilistic estimates of the impacts of various adaptation actions by expected values and other metrics given the uncertainties. If time series of actions and outcomes are included, then the dynamics of climate change and adaptation can be understood. The complete analysis allows decision makers to understand the range and timing of possible actions, and their associated costs. Its value in adaptation planning is that decision makers can search for adaptation actions that function well no matter what future climates are.

Proceedings of the First National Expert and Stakeholder Workshop on Water Infrastructure Sustainability and Adaptation to Climate Change

61

This framework was demonstratively applied to the water supply system of metropolitan Boston, MA, which consists of a large regional system serving most of the population either fully or partially and smaller, totally self-sufficient municipal water supply systems. The Climate's Long-Term Impacts on Metro Boston (CLIMB) project uses a scenario-based risk assessment approach, with a Monte Carlo simulation of impacts that uses boot strapping for climate inputs and simulation models for system performance over time. Using the indicators of reliability as measured by the percent of time it fully meets its annual demand, vulnerability as measured by size of average failure, and resilience as measured by time to recover to full operation, the self-supplied systems were found to be at risk after 2020. Several adaptation scenarios were used, including a base case, Ride it Out (RIO), Green, and Build Your Way Out (BYWO). Resiliencies of the supply system, measured by the length of a period of failure under all scenarios, are generally always one year. Vulnerabilities of the supply system, measured by the average size of deficit, is significantly less in the BYWO scenario, next followed by the Green scenario, and then the RIO scenario. Due to its present high safe yield and relatively low demand, the regional system functions well even under climate change and if the local self-supplied systems join. The regional MWRA system functions well even under climate change and if the local self-supplied systems join. Presently, the MWRA is not obligated to serve self-supplied systems in the event of temporary or permanent shortages. Therefore, local systems could consider anticipating climate and demographic changes by managing demand to minimize shortages, increasing instream flows, increasing system storage capacity though reservoirs or aquifer use, or considering using such water supply sources as reclaimed wastewater and desalination.

Click here to view Dr. Kirshen's presentation.

Click here to read the transcript of Dr. Kirshen's remarks.

Summary of Discussion Session

A water resource manager stated that the steps he has taken involve strategic planning efforts, then capital planning, then capital funding, and finally voter approval. Convincing people is very important. Most utilities have capital plans that are very rigid. It is now important to have some flexibility, particularly since adaptation needs to be built into capital plans. A local water manager responded that strategic planning is an excellent tool to communicate to both employees and the board of directors. We have had success going to ratepayers and explaining to the public their plans.

A water resource researcher mentioned that he is impressed on the mitigation efforts of EBMUD, particularly being registered with the climate registry. California requirements are driving flexibility for utilities that may not be available in other areas. A local water manager responded that AB32 was the writing on the wall for us to make the decision to join the CCAR.

A water resource researcher mentioned that with strategic planning, many issues will be affected by climate change, and utilities have to be cognizant of these other issues, such as population growth forecasts. Growth forecasts are consistently applied in strategic planning, but agreement on the forecasts is needed. Is anybody looking at a carrying capacity approach for factoring in climate change? You may arrive at the maximum amount of water that you can deliver. A local water manager remarked that they have not used it as a limiting factor, but they are integrating vulnerabilities to climate change into strategic plans. When they add climate change into the mix, they have not said that they cannot accommodate added capacity.

Another water resource researcher stated that with the issue of CSOs in Ohio, green infrastructure

Proceedings of the First National Expert and Stakeholder Workshop on
Water Infrastructure Sustainability and Adaptation to Climate Change

62

could be a good way to deal with these issues. A local water manager responded that from the regulatory agencies and the 10-15 strategic plans on water from governments, there is no appreciation for looking at future plans. It is important for people to advocate for state agencies to develop strategic planning that incorporate such factors as green infrastructure into their plans. A water resources researcher added that Cambridge, MA, is turning to LID. It is vital to start thinking about these issues now. A water resource manager remarked that Chicago formerly took a sewer oriented approach and now they focus on managing landscapes, not just pipes (Chicago has over 5,000 miles of pipes). They changed ordinances from prescriptive to performance oriented. Managing CSOs was a high level decision, and they are changing standards as they learn.

A participant stated that they are thinking of capturing non-potable water for re-use. How should they be dealing with water quality standards, total maximum daily loads (TMDLs), and state regulators? A water resources manager responded that he does not know if state agencies are thinking about these issues, but this is an issue that should be presented at American Water Works Association (AWWA) conferences, and data that show which direction the world is moving. Bringing issues to the attention of state agencies now is very important. Also use EPA to advocate your position.

A water resource manager remarked that it comes back to educating and communication. They are upgrading a plant because of algae, which led to a rate increase, and now have to describe this to their customers. The customer base has no idea what it takes to deliver clean water.

A water utility representative added that the Marcellus Shale deposit in Pennsylvania uses a lot of water to extract natural gas. Integrating planning involves so many different areas, this project hit in three weeks, and the commission was left scrambling. Many are not doing a very good job of scenario planning with high energy prices.

A water resource researcher mentioned that the Natural Resources Defense Council (NRDC) California office commented on AB 32 and looked at LID and water reuse issues. They are also looking at the stimulus and transportation bills, state revolving funds, and eventually the climate bills in order to find money for utilities. A supplementary GHG bill could allow for the sale of credits for nutrient reduction and other plans.

A member of the water resource research community stated that they are incorporating future environmental scenarios into planning, such as Everglades restoration. The key issue they are facing is sea-level rise and its affects on the Everglades. They have to determine and then evaluate the freshwater requirements. Climate change impacts all different levels of planning and it is necessary to have some event that occurs to influence decision makers. What event will occur that will make all this possible? A public referendum will not be prepared until after the fact.

A water resource manager added that ratepayers need to know more about the water process. Customer education in his community does not get funding. In contrast, Chicago looks at cultural indicators, and will not succeed in getting projects funded if they do not educate the public. It would be helpful to know what other utilities are doing with cultural indicators. A water resource manager responded that they are producing educational flyers.

Proceedings of the First National Expert and Stakeholder Workshop on
Water Infrastructure Sustainability and Adaptation to Climate Change

63

6.4   Adaptation Practices and Tools – Part II

This session illustrates a number of sophisticated planning approaches that feature the integration of multiple attributes of water resource planning within a planning framework. The presentations also demonstrated how such planning frameworks become more challenging as the multiple dimensions of future climate variability are factored into the analysis. Many ongoing research efforts are highlighted that are very much in step with this greater level of challenge.

Click here to read the transcript of the remarks of the moderator (Mikaela Engert, City of Keene, New Hampshire).

Integrated Water Management for Sustainable Water Supply in Southwestern Florida under Global Changes: Water Reuse and Energy Considerations
Mark Simpson, Manatee County Manatee County Utilities Department

Mark Simpson began by stating that historically, about 90 percent of Floridians utilize groundwater as their potable supply. Most of the population is located on the coasts. Florida has various aquifers throughout the state, including the Biscayne Aquifer and south and east aquifers of Lake Okeechobee. There is also a surficial aquifer that is generally found throughout the state with variable productivity which is recharged by local rainfall. The Floridan Aquifer is a highly productive aquifer, and well confined in southwest Florida. Manatee County is located on the southwest coast of Florida just south of Tampa Bay, and serves about 300,000 people with water and about 200,000 with wastewater. The surficial aquifer varies in depth throughout the area. There is a well-confined layer where the intermediate aquifer is found, and saltwater is found below the layer across the entire State of Florida. There is a bubble of fresh water sitting on top of the saltwater. In Manatee County, there is good confining layer between the upper Florida aquifer, the intermediate aquifer, and the surficial aquifer.

Though variations exist depending on location and the particular aquifer, water quantity, quality, and availability of groundwater often presented the most economical option for individuals and public supply, agriculture, and industry. A well that could supply all the water demands could be installed on most any piece of property. Groundwater quality is characteristically consistent, relatively easy to treat to potable standards, and seemingly unaffected by surface activities. However, increased population has led to the over pumping of aquifers. Saltwater intrusion is prevalent in aquifers right along the coasts and, as well fields have moved inland, they have led to more intrusion. The southwest Florida Water Management District has limited the pumpage on the Floridan Aquifer. They are now forced to consider alternative or non-fresh groundwater supplies, such as surface water, brackish water, desalinization, conservation, and reuse.

Surface water seems to be an ideal candidate for alternative water in Florida. There is an average rainfall of over 50 inches a year; however, only 2 percent of that precipitation goes to recharge the aquifer. The majority of the precipitation flows out to the sea or is lost to evapotranspiration. One of the other limitations of surface water quantity is that it is subject to drought, and because of the relatively flat topography in Florida, there must be large reservoirs, storage facilities, capture stations, and pump stations with transfer lines to capture the water. The cost of building these reservoirs, which have to be off-stream reservoirs, is very high. Recent studies in Peace River Valley of some new alternative service water supplies have shown $170 million to $320 million in capital costs for a 10-million gallon a day supply, which brings the unit cost up to between $4.70 and $7.80 per thousand gallons. Besides surface water quantity, it can be an issue of quality as well. Water is

Proceedings of the First National Expert and Stakeholder Workshop on
Water Infrastructure Sustainability and Adaptation to Climate Change

64

more vulnerable to contamination events and there are ideal conditions for cyanobacterial growth in Florida, elevated total organic carbon (TOC), algal exudates, and micro-pollutants. These all require advanced treatment beyond conventional treatment, which helps drive that treatment cost up.

Climate change is predicted to affect both quantity and quality of surface water. Rainfall is predicted to become "flashier" and storage becomes an even greater issue. Effects are seen sooner on surface water than on groundwater. Another alternative source is brackish water. Much of the water requires membrane treatment for total dissolved solids (TDS) reduction, which leads to high power costs that rise with increasing fuel prices. Another source, desalination, requires even tighter membranes and higher pressures and comes with the same cost effects as brackish water. Some economies can be realized with the co-location of desalination with power plants. Conservation is also identified as an alternative supply. In southwest Florida, there have been water restrictions since the mid-1990s. There are also various conservation initiatives, including education programs and landscaping. Florida has managed to decrease per capita use by about 30 percent.

Reuse has been used in southwest Florida for urban irrigation (in St. Petersburg), and throughout southwest Florida for agriculture, athletic fields, and golf courses. The Manatee Agricultural Reuse System is a local, state, and federally funded project that constructs pumping facilities and pipelines to connect three regional water reclamation facilities. The reused water is provided to farmers in place of higher quality groundwater from the Floridan aquifer. It is very cost-effective to be able to give this reused water to farmers. Very little infrastructure is required and the systems uses between 10 and 18 million gallons per year.

Mr. Simpson concluded that sustainability in southwest Florida requires multiple sources, which include fresh groundwater, the easiest source to treat and supply. The preservation of those groundwater aquifers is key for surviving drought conditions for the area. Using reclaimed water for irrigation is the cornerstone for the reuse strategy. It presents an opportunity to conserve both money and energy. Reuse presents opportunities to conserve money and energy. It is 13 times less costly than alternative surface water processes and 10.3 times less costly than brackish water processes.

Click here to view Mr. Simpson's presentation.

Click here to read the transcript of Mr. Simpson's remarks.

EPA Water Resource Adaptation Program (WRAP) Research and Development Activities on Adaptation Methods and Techniques
Roy Haught, EPA National Risk Management Research Laboratory

Roy Haught began by commenting that regardless of what we do, some degree of future climate change will occur. Climate change is affecting the global water cycle and adapting will be necessary in certain regions and for certain socioeconomic and environmental systems. In the United States, the majority of the existing drinking water and wastewater treatment plants, and water delivery distribution and wastewater sewerage collection infrastructure systems continue to age and deteriorate. Environmental and climate changes are expected to superimpose additional effects on the already stressed and aging infrastructure. The way we design, manage, operate, and maintain our water resources and infrastructure needs to be a priority. ORD is conducting research on drinking water and water quality related issues, and demonstrating how the research could be used

Proceedings of the First National Expert and Stakeholder Workshop on
Water Infrastructure Sustainability and Adaptation to Climate Change

65

to support both the aging water infrastructure and global climate change initiatives. Users can visit these R&D facilities and learn firsthand about the technologies under development.

Utilities are changing their attitudes about water usage. More stress is being placed on conservation, but water quality and quantity are still critical. The Water Resources Adaptation Program (WRAP) is conducting a drinking water treatment process evaluation. These evaluations include water availability forecasting, alternative water resource development, water conservation techniques, and water resource impacts from bio-fuel productions. ORD is developing technologies that support the WRAP program; national and regional assessments are ongoing. Results of the R&D effort will be available to end users and stakeholders. There are several facilities where R&D activities are taking place. At the Edison facility in Edison, NJ, they are taking a broad approach to infrastructure. They have developed an experimental stream facility in Milford, OH. Another facility sits on the grounds of the Cincinnati, OH wastewater treatment plant, where many different tests and evaluations can be completed. These tests and evaluations include secondary treatment on activated sludge systems, soil column for irrigation simulations, and wastewater reuse experimental testing. They are also developing a low-pressure Membrane Bioreactor Reactor (MBR) to evaluate wastewater reuse. At the facility, they can test under controlled conditions how different contaminants affect water quality.

For water conservation techniques, ORD is developing and evaluating a nonintrusive networked acoustic water quality sensors detection system to detect leaks and reduce water loss in underground pipes, prevent water quality deterioration (due to infiltration and cross contamination) in pipes, and maintain or improve water quality. Other activities include an improvement of operations management and preventive maintenance, and the development of an infrastructure database on pipe failure modes, geographic distribution, network age, and network operations. Water conservation improvements are also utilizing studies on water pipes, which include distribution system simulators such as corrosion studies, monitoring sensor studies, and leak detection studies. Alternative water conservation techniques at ORD include stormwater collection and management, rain gardens and swales development.

Mr. Haught concluded by stating that EPA ORD has research facilities, engineers, and scientists conducting research supporting 21st century sustainable infrastructure initiatives. Different infrastructure systems in the various regions of the country will depend not only on the integrity of the infrastructure, but also on the infrastructure's management, operational, and functional systems' ability to adapt to climate change. These research results could be used to help infrastructure adapt to climate change.

Click here to view Mr. Haught's presentation.

Click here to read the transcript of Mr. Haught's remarks.

BASINS CAT, WEPPCAT, and ICLUS: Modeling Tools for Assessing Watershed Sensitivity to Climate and Land Use Change
Dr. Tom Johnson, EPA National Center for Environmental Assessment

Tom Johnson began by stating that the warming of the climate system is unequivocal as is evident from the 1.4°F increase in global average air and ocean temperatures in the last century. At the same time, many regions have experienced changes in precipitation amount, an increase in the frequency of heavy precipitation events, and widespread melting of snow and ice. There has been a

Proceedings of the First National Expert and Stakeholder Workshop on
Water Infrastructure Sustainability and Adaptation to Climate Change

66

rising global average sea level since 1961 of around 1.8 mm/yr. Projecting forward, continued warming and changes in the amount, form, and intensity of precipitation are expected, albeit with large and poorly understood regional variations. Water resources and aquatic ecosystems are highly vulnerable to these changes, with possible effects including increased occurrence of floods, droughts, and water quality and ecosystem degradation. Water and watershed systems are highly vulnerable to these changes.

Climate models have limited skill predicting future climate at the spatial scales needed by water managers, particularly at local and regional scales. Models are very effective for understanding system sensitivities behavior. Looking ahead at different scenarios and drivers can help show impacts on watersheds. We can use this understanding to start identifying the ranges of potential climate change impacts and to develop strategies for managing risk (e.g., increasing resilience to future change). Given the inherent uncertainty, it is best to think about adaptation in the context of specific vulnerabilities (bottom-up vulnerability assessment). Specific systems such as wastewater plants and drinking water plants drive the question of vulnerability. It is important to constrain uncertainty and isolate what causes vulnerability. There are three modeling tools developed by the EPA ORD for assessing the sensitivity of water and watershed endpoints to climate and land use change: the BASINS Climate Assessment Tool (CAT), the WEPP Climate Assessment Tool (WEPPCAT), and the Integrated Climate and Land Use Scenarios (ICLUS) tool.

BASINS CAT is available in EPA's BASINS 4 modeling system, and provides BASINS users the capability to create scenarios and assess watershed sensitivity to climate change using the Hydrologic Simulation Program-Fortran (HSPF) watershed model and post-processing capabilities for calculating a range of management targets (endpoints) useful to water and watershed managers from model output. This capability is intended to support BASINS users interested in assessing a wide range of "what if" questions about how weather and climate could affect their systems. The system couples data and tools to support the total maximum daily load (TMDL) program by calculating permitting points, and can also assess and provide weather data scenarios. The model has a pre-processing capability to modify historical temperature and precipitation time series to create scenarios reflecting a wide range of potential changes in climate (user determined). It also has a post-processing capability to calculate hydrologic and water quality endpoints (e.g., mean flow, 100-year flood, annual nitrogen (N) loading) and can conduct frequency analysis.

It manages and automates input into the BASINS HSPF watershed model. Combined with the existing capabilities of BASINS models for assessing the impacts of land use change and management practices, the climate assessment capabilities provided by the CAT allow BASINS users to assess the impacts of alternative futures including climate and land use change as well as implementation of adaptation strategies (e.g., best management practices, BMPs) for increasing resilience to climate change. BASINS 4, including the CAT tool, can be downloaded from: http://www.epa.gov/waterscience/basins/.

WEPPCAT is an online tool that provides a similar, flexible capability for creating user-determined climate change scenarios for assessing the potential impacts of climate change on the sensitivity of soil erosion and sediment best management practices (BMPs) using the U.S. Department of Agriculture's (USDA's) Water Erosion Prediction Project (WEPP) Model. In combination with the existing capabilities of WEPP for assessing the effectiveness of management practices, WEPPCAT also can evaluate the effectiveness of strategies for managing the impacts of climate change. It has the pre-processing capability to create climate change scenarios for temperature and precipitation using the Cligen weather generator. It also has the capability for representing agricultural BMPs including grass and forested riparian buffers. The model manages input to WEPP hillslope-scale soil erosion model. WEPPCAT was developed through an interagency agreement with the USDA ARS

Proceedings of the First National Expert and Stakeholder Workshop on
Water Infrastructure Sustainability and Adaptation to Climate Change

67

Southwest Watershed Research Center and is available at:
http://typhoon.tucson.ars.ag.gov/weppcat/index.php.

The ICLUS tool is an ArcGIS application that provides a capability to derive benchmark land use change scenarios for housing density and impervious cover for the conterminous United States consistent with the IPCC Special Report on Emissions Scenarios (SRES) storylines. ICLUS outputs are derived from a demographic model and a spatial allocation model that distributes the population as housing units across the landscape for the four main SRES storylines and a base case. It uses the SERGoM cohort-component model with a gravity migration model, which includes data on county-level demography, housing density, and impervious cover. The benchmark scenarios developed for IPCC SRES: A1, A2, B1, and B2 for each decade from 2000 to 2100, allows users to run new scenarios reflecting different assumptions about population growth and density of development. Dr. Johnson concluded that although subject to uncertainty, we know enough about future climate change to start identifying the range of potential impacts and, if necessary, developing strategies for managing risk.

Click here to view Dr. Johnson's presentation.

Click here to read the transcript of Dr. Johnson's remarks.

Metropolitan Water Availability Forecasting Methods and Applications in South Florida
Dr. Ni-Bin Chang, P.E., University of Central Florida, NRMRL-UC WRAP Team

The availability of adequate fresh water is fundamental to the sustainable management of water infrastructures that support both urban needs and agricultural uses in human society. Recent drought events in the United States have threatened drinking water supplies for communities in the Chesapeake Bay area in Maryland from 2001 through September 2002, Lake Mead in Las Vegas from 2000 through 2004, the Peace River and Lake Okeechobee in South Florida in 2006, and Lake Lanier in Atlanta, Georgia, in 2007. There is a renewed interest to develop a water availability forecasting platform that serves for short-term water availability assessment and long-term water availability forecasting for large metropolitan regions. This quantitative information is critical to assist water planning agencies and utilities in water supply planning, operations, and adaptation (POA) to climate changes.

Existing drought indices include meteorology-based drought indices, such as the Palmer Drought Severity Index (PDSI). There are satellite-derived drought indices, such as the modified perpendicular drought index (MPDI) and the Keetch-Byram Drought Index (KBDI). The weaknesses of these indices include coarse spatial and temporal resolutions and no water quality information. The ideal future drought index would use multi-scale sensing and monitoring. It would contain a spatial and temporal GIS analysis of water supply availability, future supply-demand imbalance, and impacts on water quality and ecological systems. It would have remote sensing and satellite imagery available for spatial assessment of drinking water source quality and quantity, and evaluation of program effectiveness and outcomes. Finally, it would contain water utility infrastructure conditions and SDWA compliance assessment under predicted future global change scenarios (climate, demographic, and economic).

Similar to the drought ultraviolet (UV) index and air quality indices that have been widely used, the metropolitan water availability index (MWAI) presents a near real-time, risk-informed and forward-looking instrumental message in terms of both the quantity and quality of available fresh water in

Proceedings of the First National Expert and Stakeholder Workshop on
Water Infrastructure Sustainability and Adaptation to Climate Change

68

major metropolitan regions. The forecasting platform has two imbedded components. The first component is the use of multi-scale and multi-dimensional databases of optical and microwave satellite images, such as the NASA GOES, MODIS Terra and Aqua, etc., and ground-based radar stations, such as the NOAA NEXRAD system. It is intended to provide short-term (days to weeks) water availability forecasting in the form of a MWAI. The second component is the long-term precipitation periodicity analysis with assistance from hydroclimatic GCM/RCM modeling. This effort incorporates the hydroclimatic modeling with integrated ground-based sensor networks and remote sensing technologies, and aims at short-term to long-term forecasts. It is designed to forecast the future trend. Different from the existing methods, MWAI forecasting uses decision science and artificial intelligence, and incorporates both water quantity and quality information in a simple numerical range from -1 to 1.

The MWAI must be able to reflect various sources of water quantity and/or water contamination conditions in a water supply system. The index should not have any seasonality (i.e., the index should be able to indicate a drought and/or contaminant event irrespective of seasons). The index should consider water sources from reclaimed wastewater and stormwater reuse. The MWAI index should be spatially comparable, irrespective of climatic zones (humid or arid). Ni-Bin Chang concluded that the Tampa Bay, FL region was chosen in part because of its fast economic development, fast population growth, global climate change impact, and multiple sources of water. The case study demonstrates that MWAI can reflect the water quantity and quality collectively without having seasonality impact. The MWAI can account for the site-specific features from city to city region wide.

Click here to view Dr. Chang's presentation.

Click here to read the transcript of Dr. Chang's remarks.

Summary of Discussion Session

A member of the water management community offered some suggestions to EPA on siting of wastewater plants. It is difficult to find locations for wastewater plants due to the smell and pollution. They should be considered an asset for their treatment and recovery; we currently undervalue wastewater and should work on the many benefits of wastewater.

A member of the water management community wanted to know about aquifer storage and recovery (ASR). A local water manager responded that ASR is not being done in Florida on wastewater, but it is possible on potable water, up to 10 mgd.

A water manager asked about wastewater reuse, and if its demand is consistent or seasonal. And if it is seasonal, is there the ability to store the water? A local water manager responded that they have excess capacity in a surface water reservoir. By connecting all three sources, they can improve reliability. But storm surges can cause a substantial problem and they have to turn to deep well injections.

A member of the water research community commented that new indices are needed to tie in to the climate change discussion. Quality parameters have units that are difficult to add together. How do we add quality and quantity together? Dr. Chang answered that a formula designed to normalize terms (0-1) so they are all comparable would do this.

Proceedings of the First National Expert and Stakeholder Workshop on
Water Infrastructure Sustainability and Adaptation to Climate Change

69

A local water manager asked how we model for local government decision makers. Who are these tools geared to? A water manager answered that these models are open to anyone. It is important to work with small communities and make the tools very simple to use. A water researcher added that the HDPS model is very tricky to use and there is a learning curve with these models. A local water manager added that we must keep in mind who these end users are.

A water manager asked what is the experience engaging these conversations at this level with the gatekeepers of dialogue around issues such as water quality and stream flow? A water manager answered that they have engaged activist communities and used them to take ownership of the issues. An example is the Friends of the Chicago River. They paid them to do inspections of CSO, and while there was some resistance, it turned out to be an effective way to engage the community.

A water researcher commented that it takes a long time for people to accept that the government is here to help. We have set up remote sensors in small communities in Puerto Rico to monitor water quality. We established relationships and now organize the local community to take ownership of the watershed. They protected the watershed with warning signs near the wastewater treatment plant.

A water manager asked, "How do we use models to communicate with stakeholders?" OASIS is an interactive tool that can show a group of people how a community can use a limited water resource. It was a dramatic experience that enabled people to see the consequences of their actions throughout the community. To what extent do the people question the model itself?

Another water manager asked if we need new tools for wells, such as groundwater ASR. For storage and recharge, what tools are available now? A water manager responded that this illustrates the importance of front-end work. Florida already stores reused and recycled water.

A water consultant added that ASR projects are all unique and there is no general tool available that can be applied to varied situations. Modeling is a great learning exercise, but we must make sure that there is training and education when moving models out to the general public.

Proceedings of the First National Expert and Stakeholder Workshop on
Water Infrastructure Sustainability and Adaptation to Climate Change

70

7.   Moving Forward in Adaptation

Mike Shapiro, EPA Office of Water, and Pai-Yei Whung, EPA Office of the Science Advisor, provided insights on moving forward in adaptation at the concluding session. That same afternoon, participants were divided into four breakout sessions to discuss the needed techniques, tools, and research as they relate to the two workshop tracks. Two breakout groups comprised of participants in the Climate Change Impacts on Hydrology and Water Resources Management track were moderated by Karen Metchis, EPA Office of Water, and Joel Smith, Stratus Consulting, Inc. Two breakout groups comprised of participants in the Adaptive Management and Engineering, Information and Tools track were moderated by Jim Goodrich and Jeff Yang, EPA Office of Research and Development, Elizabeth Corr, EPA Office of Water, and John Cromwell, Stratus Consulting, Inc. A subsequent discussion was held with a cross-section of participants from both tracks to reflect on the workshop and to explore some of the concepts and needs heard during the workshop. These three sessions provided an extensive list of suggestions and ideas for moving forward in adaptation.

7.1   Concluding Remarks

Before adjourning the workshop, Dr. Pai-Yei Whung and Dr. Michael Shapiro commented on the significance of the workshop and potential opportunities and next steps that came out of the presentations and discussions, including collaboration, research, and data collection.

Dr. Pai-Yei Whung, Chief Scientist, EPA Office of the Science Advisor

Pai-Yei Whung provided concluding remarks on the significance of the discussions at this workshop in the context of EPA's overall approach to building sound science and technology. Dr. Whung highlighted the importance of ensuring that the real costs of water are considered in decision making, and that financing is a key element of water resource planning. If we consider financing in the early stages of our planning processes, we will have more beneficial outcomes.

Dr. Whung provided a brief overview of EPA's activities under its national water program. The goals of the program include:

1. Use core water programs to contribute to GHG mitigation.

2. Adapt implementation of core water programs to maintain and improve program effectiveness in the context of a changing climate and assist states and communities in this effort.

3. Strengthen the link between EPA water programs and climate change research.

4. Educate water program professionals and stakeholders on climate change impacts on water resources and water programs.

5. Establish the management capability within the National Water Program to engage climate change challenges on a sustained basis.

Dr. Whung explained that the activities in support of decision making need to consider two key technical factors: incorporating full cost accounting and lifecycle analysis of water used for energy development, and using climate information gathered through coordinated and comprehensive observations to inform decision making. The Global Earth Observation System of Systems (GEOSS), which is a platform for bringing together integrated climate observations to meet user needs, is an example of a resource that can be used to make informed decisions in the face of climate changes.

Dr. Whung addressed the importance of building collaborative relationships between the science and technology communities, emphasizing the importance of leveraging resources from the private and public sectors and academia. In addition, in water management planning, there needs to be explicit inclusion of climate change adaptation considerations. Dr. Whung also stressed the importance of monitoring and modeling priorities, such as strengthening the data network, and creating a water data portal. Lastly, she highlighted the importance of integrating science and technology in clear policy actions, and for incorporating into these policy actions feedback from practitioners on what information they need from the science community.

Dr. Whung concluded by reminding the group that this stakeholder workshop is a first step, and that there will be continuing action. EPA is making these activities a priority for the Science Policy Council Subcommittee for Agency Science Priorities.

Click here to view Dr. Whung's presentation.

Dr. Michael Shapiro, Deputy Assistant Administrator, EPA Office of Water

Michael Shapiro made the observation that despite the many uncertainties, he was extremely impressed with the amount of work that has already been done on these issues. We are moving forward simply by acknowledging the problems. However, we cannot wait for the perfect answers to our questions. We need the research to move forward toward policy immediately. We are ready to build upon the lessons learned through current activities. Climate change has caused us to reflect back on other hidden needs in the water world. We need a suite of tools to help us make decisions, both in the short-term and long-term. Figuring out how to proceed from here will be difficult. Immediate steps that EPA plans to take to continue the momentum built at this workshop include:

- Posting presentations on the EPA WRAP Web Site.

- Making proceedings from the workshop available.

- Follow up this workshop by evaluating the information generated in the discussions from these past two days to identify actions that can be pursued in the next phase.

Mr. Shapiro explained that the products of this workshop will help EPA revisit the strategy it set forth in its National Water Program Strategy Response to Climate Change, and will allow EPA to leverage its resources in the most effective manner.

7.1   Suggested Ideas and Recommendations for Moving Forward in Adaptation

Participants in the breakout sessions on climate change impacts on hydrology and water resource management were asked to comment on the following three questions:

- What techniques and tools do water utility managers and engineers need from the research community for decision making and vulnerability assessments?

Proceedings of the First National Expert and Stakeholder Workshop on
Water Infrastructure Sustainability and Adaptation to Climate Change

72

- What information is available now or can readily be made available to support water utility decision making and adaptation efforts?
- Where are the gaps that need to be filled to enable water utilities to undertake adaptation actions?

For the breakout sessions on adaptive management and engineering, participants were asked to comment on the following three questions:

- What techniques, tools, and information are needed for decision making would utility decision makers and engineers like from the research community?
- What policy relevant information can the research community provide now or in the near term?
- Given what is known, unknown, and likely/possible to be known in the near term, how should decisions on infrastructure be made?

During the small group work session, participants were asked to provide recommended ideas relative to research, tools, and information in order for EPA to:

- Gather individual perspectives and understanding of who is doing what and where gaps and challenges exist.
- Provide individual feedback on the research activities of the EPA Office of Research and Development (ORD) as presented during the workshop in order to inform EPA's future research directions.
- Assist EPA and other federal agencies and research organizations in evaluating opportunities to help meet the needs of water and wastewater utilities.
- Identify possible next steps in developing tools, projects, or programs that could help water and wastewater infrastructure managers prepare to adapt to climate change in the near term.

Participants in the breakout sessions and the post-workshop discussion provided numerous suggestions, ideas, and comments on models, data, decision tools, uncertainty, case studies, and other issues, as identified below. Those ideas that appear in *italics* were those that were suggested during the half-day work session that followed the workshop.

Participants in the breakout sessions on climate change impacts on hydrology and water resource management were asked to comment on the following three questions:

- What techniques and tools do water utility managers and engineers need from the research community for decision making and vulnerability assessments?
- What information is available now or can readily be made available to support water utility decision making and adaptation efforts?
- Where are the gaps that need to be filled to enable water utilities to undertake adaptation actions?

For the breakout sessions on adaptive management and engineering, participants were asked to comment on the following three questions:

- What techniques, tools, and information are needed for decision making would utility decision makers and engineers like from the research community?

- What policy relevant information can the research community provide now or in the near term?

- Given what is known, unknown, and likely/possible to be known in the near term, how should decisions on infrastructure be made?

Participants in the breakout sessions and the post-workshop discussion provided numerous suggestions, ideas, and comments on models, data, decision tools, uncertainty, case studies, and other issues, as identified below. The points raised are organized by themes and topics within these themes and topics:

- Difficulty of adaptation under uncertainty about climate change

- Information needed by water managers, including:

    - Information relating to current hydrology and climate, including paleoclimate, hydrometeorological information needs, and reference data.

    - Information relating to climate change projections, including GCM downscaling archive and standards, model outputs and actionable science, and model inputs and the use of models.

- Adaptations in engineering practices and decision making, including:

    - Use of climate change information in decision-making, risk management, and vulnerability assessment,

    - Engineering practice,

    - Multi-factor hydro modeling, and

    - Economic tools.

- Technologies and Management, including:

    - Drinking water supply and demand management, and

    - Reuse and aquifer storage and recovery.

- Clearinghouse that includes and addresses:

    - Case studies and best management practices,

    - Tools and technical assistance,

    - Regional and local information and tools, and

    - Special needs of small utilities.

- Federal role and interactions with federal agencies.

- The role of EPA, including implications to water quality programs.

- Water, energy, and greenhouse gas emissions.

- Communication, education, and public outreach.

Those ideas that appear in italics were those that were suggested during the half-day work session that followed the workshop.

Proceedings of the First National Expert and Stakeholder Workshop on
Water Infrastructure Sustainability and Adaptation to Climate Change

74

Difficulty of Adaptation under Uncertainty about Climate Change

- In the workshop, two paradigms emerged for looking at the nexus of climate science management and water engineering. The Lettenmaier – Behar – Estes-Smargiassi paradigm holds that engineers and utilities know the cost-benefit analysis of different adaptation options and need probabilistic forecasts to determine which approach to take. The Brown – Waage paradigm holds that there is irreducible uncertainty in all forecasting, and the decision processes and tools need to be changed.

- There is a failure in that uncertainty is not included explicitly in many models.

- On the idea of a "cone of uncertainty," Dr. Lettenmaier used to throw out some scenarios, but now keeps all of them within the cone. Some of the scenarios might be less credible than others.

- With more models, there is a greater range, but it is a mathematical lowering of the lower bound. With more models, we will have more models that are credible. For example, El Niño is included in some models, with a lot of uncertainty.

- We do not know enough from the climate science community to be moving ahead with much certainty. There is an immediate need for greater certainty with respect to precipitation forecasts. The climate community needs to stop providing too much information that is lacking in specifics.

- There needs to be more focused research on decision-making under uncertainty in the water resources management context. We need a system whereby the uncertainty in modeling can be managed appropriately. This will require alterations in decision-making protocols and different means of quantifying uncertainty (e.g., decision scaling).

- Guidance is critical for communicating uncertainty in model projections to decision-makers.

- We will not do away with uncertainty, but we need to understand it better. Research will not take away uncertainty, and utilities have discretion in light of uncertainty.

- Hydrology is already uncertain and climate change adds more uncertainty. We need to consider what lessons could be learned from how hydrologists deal with uncertainty now, as well as the implications for additional uncertainty.

- How do utilities integrate all the issues with different temporal horizons and variable levels of uncertainty? How do they decide on critical impacts to infrastructure over the long-term? Decisions are being made today, so we need this information quickly.

- We need a different model for decision-making under uncertainty. We cannot get away from uncertainty, and we cannot get hung up on it. Robust decision-making will enhance risk management, as we will be held accountable for our decisions. Uncertainty is inherent in planning.

- There is less confidence regarding uncertainty. When uncertainty is larger, we may need to use the second band of uncertainty when there is no consensus trendline.

- There are many tools out there, which need to be deployed appropriately, as well as be robust and explicitly incorporate uncertainty. We need better general information on the potential consequences, based on probability and taking into account the cost of not taking adaptation actions.

- *Uncertainty is a moving target until the Federal government sets a rule; then the utilities do not have to deal with it on their own.*

- *We are not there yet. Each state/locality must make its own determination.*
- *We are at a point of transition on how to cope with this uncertainty on a systemic basis.*

- A primer on uncertainty should be developed and utilities should have access to it. The U.S. Climate Change Science Program and the IPCC have developed such a primer.

- Utilities should have a primer on uncertainty.

- There are a number of gaps in understanding uncertainties inherent in model projections that are already available. The expert community needs to provide guidance in layman's terms for non-climatologists so they can understand what variability exists in model projections. The question is, "How do we communicate the uncertainty and pass it along to the decision-makers?" From storylines to emissions – down to models and downscaled outputs – there is much uncertainty. Guidance is critical.

Information Needed by Water Managers – Current Hydrology and Climate

Paleoclimate

- Paleoclimate reconstructions is not being used. When we look at historical variability of climate, the reconstructions are there, especially in tree rings, going back more than 500 years. If we can build systems using that variability, we will have greater resiliency.

- Model data exist and are being used, but paleoclimatic data could be better used. Data on Colorado River stream flow have changed the paradigm; general circulation model (GCM) and stream flow data should be combined.

Hydrometeorological Information Needs

- There is a need to identify and elaborate the linkages between natural flow regimes (i.e., natural water quality) and best management practices (e.g., low impact development, reforestation, afforestation). We need to determine the scale of the relationship between these factors and what critical masses for change may be necessary before one affects the other.

- Improvements on some models (such as runoff, water quality) are necessary before they can be used within new climate change models (meta-models). We can both improve models and encourage use of existing models.

- There is a need for better information about the impacts of climate change on the hydrological cycle at sufficient resolution to be actionable, and efforts toward this objective are not moving forward fast enough.

- With respect to watershed or regional planning, large-scale solutions are the focus of research. Researchers need to be engaged in regional planning. GCM data need to be reassessed in the light of stream flow and groundwater information. Environmental interests encourage focusing only on the impact of climate change, but water infrastructure research can help the environment.

- With respect to the variable infiltration capacity (VIC) and Stanford models, hydrologic modeling tends to be watershed specific, which requires energy to develop.

Proceedings of the First National Expert and Stakeholder Workshop on
Water Infrastructure Sustainability and Adaptation to Climate Change

76

- Up-to-date hydrologic and climatological data at the local and regional levels, with downscaled models, are needed.

- *Improved forecasts on precipitation and climate variability*

- GCM data need to be reassessed in the light of observed stream flow and groundwater information.

- Reasonable and accurate local and regional precipitation models are needed, particularly given the peaking factors at some plants as a result of storm events.

- We lack data on small watersheds, as there are fewer monitoring stations. Climate change is complex and overwhelming to small water systems. We need to downscale not just climate models, but operational models as well so that utilities can plan for adaptations.

- Need better indicators to track progress.

Reference Data

- As for documenting precipitation frequency, it would be interesting to consider a Bulletin 17B-equivalent for climate change. U.S. Geological Survey administers Bulletin 17B, which establishes a set of procedures for estimating flood frequency and magnitude from coast to coast, with a purpose of establishing uniformity. There is a need for guidance of this sort for practitioners, but there is also an apparent lack of data.

- Intensity-duration-frequency (IDF) curves are an example of another need.

- Updated precipitation frequency estimates are needed.

- *Probability distributions for changes in parameters, rainfall intensity, and frequency (by decade) for the next 100 years*

- There is a need for model-based probability functions for temperature and precipitation.

- *Need to narrow the uncertainty on precipitation, drought, and storm intensity, which will be used in design standards*

- Revised flood hazard data that account for climate change and imperviousness are needed.

- Utility managers should have access to an updated analysis of relative hydroclimatic variables such as 24-hour rainfall and 100-year flood. As the perspectives on these variables change, we should have updated metrics with the most recent data methods and also a usage document that discusses the potential for non-stationarity perspectives.

- *Hydrometerological characterization of risk: 7Q10, 100-yr flood, and rainfall intensity duration. Develop up-to-date methods and data, along with peer-reviewed discussions of what we know and what we think about what are the likely trends in risk characterizations at the high and low end of hydrometerological systems*

- Quantitative data are needed to show the potential impacts of climate change; otherwise it cannot be taken into account.

- There is no consensus with respect to the available data. Additional data sets are needed. Models are limited by the data available.

- There is a lot of data out there, but there is a disconnect between data in the science/academic world and the utilities. Data are needed in the short-term, downscaled modeling will come later.

Proceedings of the First National Expert and Stakeholder Workshop on
Water Infrastructure Sustainability and Adaptation to Climate Change

77

- From a local utility standpoint, there is a need for observed data sets. Some of the lowest cost adaptation measures are operational changes that require real-time data feedback. Local governments want to respond to climate variability in real-time. There is a certain amount in forecasts now, but we need to know how reliable they are.

- It would be helpful to have a tool and to run some forecasts through it, and then be able to enter observed data to test its effectiveness. A one-week lead time would be helpful, but it would be a good idea to have "now-casting" capabilities with two or three hours of lead time.

- There needs to be more information on thresholds of significance. Maybe there should be some testing to flesh out these thresholds.

- The country's efforts in monitoring environmental changes need attention. We are losing monitoring stations at a rapid pace. We need to modernize and update the environmental monitoring infrastructure, and we need to encourage comparable action in other countries. We are not just looking at water impacts, but we are primarily focused on temperature, precipitation, and stream flow.

- *Better environmental monitoring, including not losing current capability and developing new systems (satellites and surface monitoring)*

- *Ecosystem impacts and biological research (e.g., in the West, habitat/species drive water management)*

Information Needed by Water Managers – Climate Change Projections

GCM Downscaling Archive and Standards

- With the next generation of Intergovernmental Panel on Climate Change (IPCC) scenarios, more effort needs to be put into archiving data, at least for those GCMs that are based on IPCC emissions scenarios.

- There needs to be more information available regarding the collection method, results of peer review, etc. Standardized data are important.

- There are a number of gaps in understanding uncertainties inherent in model projections that are already available. The expert community needs to provide guidance in layman's terms for non-climatologists so they can understand what variability exists in model projections. The question is, "How do we communicate the uncertainty and pass it along to the decision-makers?" From storylines to emissions – down to models and downscaled outputs – there is much uncertainty. Guidance is critical.

- Many decision-makers, including those on water boards, need a translation of models. Some utilities must have the translation already. It is not that the translation does not exist. Where does the translation come from?

- There is a need for better climate information and forecasting, and the need to be rid of generalities and handicapping uncertainties. Those uncertainties that cannot be gotten rid of need to be factored into planning. Given a specified emissions scenario, utilities want to know the uncertainties in each model. There are still improvements that can be made in the tools to help narrow this variability.

- *Credential a set of forecasts and tools that have been peer reviewed, that are workable, and in which there is some confidence. Key questions include:*

Proceedings of the First National Expert and Stakeholder Workshop on
Water Infrastructure Sustainability and Adaptation to Climate Change

78

- *What data do you use?*
- *What are the directions of the changes?*
- *What is the timing?*

- Utilities should know the progress and expected future progress of the precision and accuracy of GCMs so that they may know whether to make a decision now based on existing models or wait a given amount of time for certainty to improve.

- *Do users know what to do with the downscaling? There is confusion. How do you work with the information you have? How do you make decisions?*

- *At the OSTP meeting, it was discussed that the next general circulation models will be at the 50 km grid size. In the next five months, they will want water resource input.*
  - *We need to tell modelers what decisions are being made, and what outputs are needed.*
  - *Downscaling by objective*
  - *NOAA did not really talk with water resource agencies*

- *Given uncertainty of climate model scale, NAS needs to conduct a study on the state of science and practices.*

Model Outputs and Actionable Science

- *What data are actually tangible, and what science is actionable?*

- When receiving outputs from models and research, we need to know about their inputs. How certain is the projection, and can it be used in formulating new projections? What is the certainty of the output? We need to know the quality, certainty, assumptions, and appropriateness of the data.

- *At the NYC DEP, a climatologist is always in the room to interpret information to engineers, person-to-person. How can we make that happen more often?*

- *Better mechanism to provide climate science expertise to users of downscaled products (i.e., how to use and interpret models). This is possibly a role for a national climate service.*

- *What research is available that is valid? In what ways can it be used?*

- There is a need for standardization of downscaled model output, looking toward the next generation of statistical downscaling.

- Perfect models are desired at the expense of using good, but not perfect models. The question is whether a model can provide output that is good enough to be actionable.

- Having the tools to process the model outputs is important for use in hydrologic models. Many municipal utilities have models set up that could use the GCM outputs, but they do not know how to do it. Guidance on what models to run, and what input to use, would be very useful.

- *Identify the relationship of climate parameters and model outputs to impacts and geographic differentiation*

Proceedings of the First National Expert and Stakeholder Workshop on
Water Infrastructure Sustainability and Adaptation to Climate Change

79

Model Inputs & Use of Models

- When multiple models are available, often the approach is to choose a single model. There are approaches, such as Bayesian models, that use several models simultaneously.

- Engineers do not want to use an ensemble of models.

- A set of accepted guidelines is needed for creating scenarios for model inputs.

- There is a significant amount of redundancy in existing databases. Tools are needed to simplify the process of importing data into the databases to make the data directly and easily accessible for watershed modeling purposes.

- The stakeholder process is very much affected by assumptions. The models often do not give an answer, or inform us about uncertainty. How do we get the process to the stakeholder level (i.e., going through the model with the stakeholder)? Stakeholders will engage and understand if they can vary inputs and learn how inputs change. We have observed this using the OASIS tool.

Adaptations in Engineering Practices and Decision Making

Use of Climate Change Information in Decision-Making / Risk Management / Vulnerability Assessment

- *Need information to help more sophisticated utilities with decision making and risk management in the face of uncertainty*

- Policy-makers need more research than utility managers and engineers; the Chesapeake Bay was used as an example.

- There is a need to find out why people make decisions contrary to tools' recommendations. What barriers exist in the decision-making process at the human level and the institutional level? Why do ratepayers and politicians, for example, make decisions that contradict the tools?

- With the management paradigm, information should be sought on "no-regrets" or robust solutions that lead to greater flexibility and efficiency. (Do these solutions exist? How many are there? How well do they work under climate change? What methods are there for determining whether a solution is robust or no-regrets?)

- Risk management needs simple decision trees to help integrate climate change into planning.

- What utilities most need help with is developing flexible adaptation pathways. For example, if sea levels rise a half-meter by 2050, utilities might not need to build retaining walls immediately, but they need to be able to plan to leave space for a wall in the future.

- Would like to better synthesize a way to provide information on what the available options are, and how to choose among them.

- If there are 10 GCMs, and 8 predict that there will be increased water, but the *greatest risk* is on the storage side, that risk should be considered.

- We need to explore the costs of 100% reliability. People are reluctant to use a new method if it increases the risk of failure, as we have seen with seasonal climate forecasting. We need to get rid of the risk of catastrophic failure.

- *A risk management framework opens the question that something is not 100% fail safe*

Proceedings of the First National Expert and Stakeholder Workshop on
Water Infrastructure Sustainability and Adaptation to Climate Change

80

- *Risk management has always been around, and always evolves*
- *Risk management framework is part of education*

- *There is a downside of more information – costs associated with increased precision and reducing uncertainty*

- Appropriate tools and data exist for us to begin risk management projections today, including probabilistic analysis. We need a formalized approach for implementing water management risk analysis, from the water utility standpoint.

- *With respect to actionable data versus risk management frameworks, both types of tools are needed*

- With regard to the notion of having two overarching paradigms, the goal should not be to reconcile the two paradigms, but to make progress using both approaches. (The Lettenmaier – Behar – Estes-Smargiassi paradigm vs. the Brown – Waage paradigm)

- With respect to the various assessment tools that are needed, utilities are doing a lot of these kinds of assessments already. The lack of tools is not a justification for a lack of action. Instead of talking about tools that need to be developed, utilities need to have more money, the technical expertise, or a legal requirement to act. Utilities' willingness, not their ability to act, is the problem.
    - Should these be decision tools or scientific tools? Both risk and adaptation response tools exist, but they are not being applied.
    - Utilities that are doing these assessments are doing them more as experiments than as routine processes.

- Many of the techniques and decision support tools exist. Capacity building, information transfer, and support services to implement and apply the tools are needed.
    - Unsure if the tools exist for utilities because there are still information/data needs (e.g., storm intensity, precipitation).
    - Utilities cannot wait for the information and a scenario-based approach must be used.

- In applying decision tools, we are always trying to maximize flexibility and resiliency. To do this, we need shorter planning horizons. Traditionally, utilities prepare 30-year plans and make some adjustments along the way. However, the changes required could become more drastic. Utilities practice adaptive management, but they do not communicate [the decisions] well.

- *Is there a way to tier recommendations for major things? That is, long-term design decisions vs. near-term tier one decisions (e.g., those actions that are affordable, doable in near term for next summer, etc.)*

- Flexibility is expensive and compliance with the current, known environment is already expensive. Cost analysis and risk assessment tools for decision support are needed, along with some consideration of what the error might be.

- Improved tools are needed to assess whether a risk is significant to a particular utility. Benchmarks and metrics should be used, so that assessors and utilities can compare the risk with other uncertainties. In other words, we need the ability to determine whether a utility is particularly vulnerable to climate change compared to other stressors.

- *Every community will make a different decision based on local circumstances*

- *Utilities put off decisions as much as possible in hopes of being more certain, but at some point must make the decisions*

Proceedings of the First National Expert and Stakeholder Workshop on
Water Infrastructure Sustainability and Adaptation to Climate Change

81

- *Tendency is to design with as much flexibility/capability as they can afford*

- In California, there is concern that the state department of water resources will issue publications detailing projected climate change impacts and expect the local utilities to act on a state-sanctioned set of projections. The utilities believe that they are better positioned to do this themselves.

- Risk, probabilistic, or actuarial information should be used in these assessments so that we can have input on how insurance companies will use it and be able to convey the information to the public.

- Studies should be reconfigured to consider climate impacts only in the context of adaptation. The implicit objective of climate studies is to inspire climate change mitigation and to spur us into action. Therefore, they emphasize negative results. We are using the same studies, including model output, that were intended to spur us into action for the purpose of projecting how utilities must adapt to change. Instead, we need studies that are focused more on adaptation and sector-specific vulnerability, and move into climate change secondarily. As a result of the current process, we ignore too many vulnerabilities.

Engineering Practice

- Where do these small engineers get their information now?
    - 1046 training is dying out, but a new way to educate them and provide new information has not been given to them
    - There are training centers and professional associations available for training.

- *Could include information on how engineering schools incorporate the shift in stationarity, and use of resilient design. For example, University of Cincinnati has a course in "design for uncertainty."*

- Engineers should be better educated on uncertainties in order for them to overcome reluctance to use these tools.

- *It is a challenge to persuade an engineer to expand his/her awareness or comfort zone.*
    - *Rate payers will pay when risk reaches unacceptable level.*
    - *Engineers need tools to understand what they need to adapt to.*
    - *Our job is to deliver that.*

- Water systems (especially the smaller systems) are struggling with standards, particularly with how engineering standards need to be modified because of climate change. Standards also need to be defensible and hold up in court (e.g., justification for why a pipe was engineered to be larger than it normally would). Standard methodologies and/or adjustment factors for existing standards that could be widely accepted would be helpful.

- Would like to distribute information in the form of new manuals or textbooks. Managers and engineers need to know new, updated information, not necessarily in the form of case studies.

- *The big shift is away from stationarity. This issue is similar to the growth issue.*

- Stationarity is dead, and we need manuals and guidebooks for junior engineers. There is also a need to keep guidebooks up-to-date. Engineering firms provide a good design for downscaling. They require that scenarios are appropriate for regional levels.

Proceedings of the First National Expert and Stakeholder Workshop on
Water Infrastructure Sustainability and Adaptation to Climate Change

82

- Having information on the implications for the design of plants and operational protocols would be helpful. The operational response to climate change is the most important. We need to look into developing guidance to help utilities understand how variability affects infrastructure design and planning.

- *How do we change design standards to build resilience?*

- *There is uncertainty in engineering design, e.g., Milwaukee's cryptosporidium problem was caused by a storm and improper operations*

- The end of percentage-based reliability as a concept would be useful in the future assessment of adaptation. Utilities should accept that there may be extreme, outlying climate events, and they should develop a continuous cost curve to account for all years. Flood control, interconnectivity, and design for failure all need to be considered.

- There needs to be a training program for engineers and practitioners (possibly an online training program) that focuses on the mechanics of water supply impacts, so that engineers can characterize impacts. Many engineers are being tasked to do such projects but lack the necessary training to do it effectively.

- Green infrastructure's long-term capabilities need to be measured more accurately, particularly how communities might benefit. It would be good to issue a Request for Proposal (RFP) for a grey infrastructure vs. green infrastructure comparison to contrast the differences.

- It is important to make it possible for communities interested in green infrastructure to be able to get earmarked funds.

- Big utilities are well staffed and aware of issues, but these issues need to be translated to medium- and small-sized utilities. Developing best practices would be the best option, particularly with engineering.

- We have been adapting to climate change for thousands of years; we need to market increased robustness to withstand the challenges of climate change

Multi-factor Hydro Modeling

- There is a need for a broader focus on hydrologic changes as opposed to climate change. Climate change is one element of hydrologic change (in addition to changes in land use and water management structures). When you look at the trends in stream flows, the reasons are not all well understood, but we have huge signatures of land cover change. Some argue that this is a bigger issue than climate change. We need a broader focus on hydrological changes, vis-à-vis the "stationarity is dead" concept.

- There is a great deal of evidence showing that stream temperature changes significantly with land use changes.

- There are numerous feedbacks in hydrological models, and it is impossible to separate hydrological changes from land cover changes. One affects the other.

- Watershed-based models that are linked to comprehensive land-use impacts are needed. This linkage could then inform design standards. Utilities typically respond to, rather than shape, land-use decisions.

- Since they look out into the future, the models are not mindful of socioeconomic changes and issues. These issues often overwhelm water-specific issues.

Proceedings of the First National Expert and Stakeholder Workshop on
Water Infrastructure Sustainability and Adaptation to Climate Change

83

- Reliable forecasting of various parameters, including human habitability, natural systems needs, and water availability, is needed. How much water will be available to utilities considering projected population and other factors?

- Utilities should not be using models just for climate change indicators. They should also consider demographic, water quality, and sea level indicators. In Denver's decision process, only one scenario incorporates climate change. Of all these agents of change, climate change is the last of the utilities' priorities. Climate projections *alone* might not change a utility's decision process, but climate projections *in addition to* the other projections might change the decision process.

Economic Tools

- *Economical tools/approaches are needed to assess the impacts of climate change*

- Financial and economic tools should be used in an engineering approach (e.g., the future cost of action, willingness to pay by rate payers, and the relative price of current inaction and alternative approaches). Integrating economics is a very complex issue, and managers and policy-makers are making complex decisions without all the available economic information.

- A holistic financial model that allows for pricing different scenarios is needed. Utilities are good at pricing projects, but they need to begin to think more broadly to put a price on adaptation.

- Research is needed on customer attitudes, perceptions, desires, and willingness to pay for adaptation. These results could then be taken to elected officials, as their constituents are the utilities' customers.

- We price the variables of climate forecasting that we are able to price, and do not price the variables that we are not able to price. Existing climate models have precisely priced all variables that can be priced, but there are fundamental ranges of uncertainty and other unpriced items that at least need to be inventoried (examples include the El Niño-Southern Oscillation and methane evaporating from permafrost).

- If a water manager wants to value more intense events, are there cost-effective risk reduction measures that can be applied (e.g., paradigm; no-regrets strategies)? Decision-makers need to know the cost of building versus the cost of catastrophe. Water districts do not do this because the information is not available. In the case of the Water Security Act, we had no certainty that water supplies would be targeted in a terrorist attack, but we enacted it anyway

- *Need definition of the costs and economic impacts of extreme events, as well as the costs of corrections*

- We need to calculate the environmental benefits. We need to quantify the environmental benefits between green and grey infrastructure. We also need to quantify the net environmental benefit between centralized and decentralized systems.

- There are better data available about the impacts of smart growth and run-off. There is a need to better quantify the true costs of urban sprawl and communicate these costs to the public. The public does not grasp these issues of the built environment. There is no disclosure on the impacts of sprawl.

Proceedings of the First National Expert and Stakeholder Workshop on
Water Infrastructure Sustainability and Adaptation to Climate Change

84

Technologies and Management

Drinking Water Supply and Demand Management

- A national inventory of drinking water sources needs to be modeled relative to expected demographic shifts. Planners need to know if the drinking water supply will be in balance with the population. In the Great Lakes region, utilities are already seeing a decreased payer base. There is water, but no one to drink it.

- There needs to be more research on building climate change explicitly into water supply impact studies. There has been work on this topic, but we need to broaden the base of practitioner use.

- There is less uncertainty with respect to water demand, relative to the uncertainty of climate change processes.

- There are currently no good tools for estimating the levels of change in demand based on different factors. For each sector involved, we should have tools that can consider both physical and natural factors which can affect demand.

- *Document the demand-side effects of climate change. The science and data associated with the impacts of climate change on demand are pathetic. There is little interest.*

- *Need ability to forecast effects of conservation, climate, and population. The Federal government's ability to forecast demand has crumbled.*

- *Need good data on the effect of conservation*

  - *Who does the generalization? (EPA? USGS?)*

  - *Nationwide water use projections are wrong*

  - *COE used to be the powerhouse, but not anymore*

  - *Bureau of Reclamation punted on demand in a recent study, as it did not have science or methods under a climate scenario*

- *Public responds to heat by watering more versus less landscaping*

- *Institute for Water Research has developed a model, now in use by CDM*

- *There is a WRF RFP on water demand on water utilities under different climate scenarios*

- Pricing mechanisms and designing rate structure as a demand-side management tool exists. It should be disseminated to managers and political leaders.

Reuse and Aquifer Storage and Recovery

- *Need more information on conservation and water use/reuse*

- The current NAS study on water reuse is partially funded by EPA. A plan for a sustainable water supply would be wide in scope, including water reuse.

- In New York, the media dubbed water reuse as "toilet to tap" and the approach quickly lost momentum. Public response is largely driven by the media, as well as reports of pharmaceutical risks in outflows. This represents risk, which makes people uncomfortable.

- Water reuse has been successful in some places.

- New Jersey is the first state to use aquifer storage and recovery (ASR), which takes surface water and injects it into the ground to use later. In New England, there is limited information

Proceedings of the First National Expert and Stakeholder Workshop on
Water Infrastructure Sustainability and Adaptation to Climate Change

85

available for this type of injection; the quality of the water to be injected may not match the quality of the water underground.

- We are limiting ourselves environmentally by not using wastewater. Regulations and technological obstacles are in the way, but research and development (R&D) would be money well spent to try and overcome these hurdles.

- Individual managers do not know about wastewater reuse and water plant managers need more training on this subject. We need to influence these issues.

- EPA could play a leadership role with respect to sustainable water supplies. There have been improvements recently in Los Angeles County and good outreach as well. Water reuse is underappreciated as a strategy and is always considered a last resort, largely based on public perception. Research, advocacy, and leadership on water reuse are needed to ensure a sustainable water supply.

Clearinghouse

- *Establish a clearinghouse of information, which would allow for better coordination and organization. There are a lot of workshops and they are not moving things forward.*

- A national data portal of water information from all government agencies is needed. A single portal would facilitate easy public access. Interagency coordination would be critical given that this issue is broader than EPA's realm. Information is too scattered across the federal government, and there is a need to link to and access information that is housed in various agencies.

- *Need guidance on how utilities can adapt to climate change*

    - *How to use climate change model output*

    - *Review of tools that can help in adaptation*

- *Need peer-reviewed information and an understanding of how to use the information (its limits, etc.)*

- *Clearinghouse should include success stories*

- *Distribute annual bulletin summarizing critical information and studies that have come out for last year. Then, the IPCC can summarize it all every five years.*

- There should be a web-based forum for sharing case studies and ideas. This could serve as a centralized, reliable source of information. The Association of Metropolitan Water Agencies just created a catalog of federally funded publicly available information on climate change and water. They are looking for ways to continue the project, including maintaining and updating the database.

- The Water Research Foundation is creating a climate change clearinghouse to make information available to utilities in a single location.

- Approach the clearinghouse concept with caution: think about how to design it well, and make sure stakeholders do it together or it will not succeed.

- It will be a challenge to coordinate efforts (e.g., both WRF and AMWA issued clearinghouse RFPs)

Proceedings of the First National Expert and Stakeholder Workshop on
Water Infrastructure Sustainability and Adaptation to Climate Change

86

<u>*Case Studies and Best Management Practices*</u>

- Case studies of adaptation practices for utilities are needed.

- *Develop case studies*

  - *Case histories of planned development to show what works, which can then be used to explain to taxpayers why certain projects are needed*

  - *Case studies on extreme events and effects on wastewater treatment plant operations*

  - *Case studies on how much is saved by planning ahead, instead of by catching up, which will help sell adaptation (e.g., build in adaptation during the rehab cycle)*

- It would be good to start assembling case histories where projects have been implemented that are driven by climate change. This would be helpful for utilities to be able to learn from each other (e.g., with respect to how they can communicate with their ratepayers about the need for adaptation). Climate change best management practices are key.

- Case studies are typically of large utilities and written by/for university researchers, which should be more accessible.

- We should seek examples of projects and activities that have worked, based on case studies rather than a one-size-fits-all approach.

- Case studies will help illustrate many of the issues, particularly with respect to extreme events and effects on wastewater treatment operations (collection and treatment).

- Effort should be put into building success stories. For instance, you can set one aside as a demonstration site, which takes advantage of the latest technologies.

- *Conduct case studies on three large utilities that went from General Circulation Models to plans: King County, Boston, and New York City*

- It would help to get other utilities who are taking action with respect to adaptation to make presentations to decision-makers in other locales. Information from this conference should be distributed to a larger group.

- Having access to best management practices is key.

- *Best management practices for better operation of existing infrastructure, including reservoirs and wastewater infrastructure*

- *WRF work: Philadelphia and New York City had modern water systems by 1840s, but now they are aging*

  - *Articulate different issues for the old East versus the new West*

  - *Learn from each other*

  - *Explore the common elements, differences*

- *Australians/British include adaptation in their asset management, but they only use the Hadley model (that is a defect)*

Proceedings of the First National Expert and Stakeholder Workshop on
Water Infrastructure Sustainability and Adaptation to Climate Change

87

- *San Francisco's asset management model has no feedback loop to climate change*

Tools and Technical Assistance

- *Need tools and training to help local governments and utility decision makers understand uncertainties of forecasts and help them make/implement decisions (a new kind of uncertainty). Need better ways of making decisions.*
- *Modeling community talks to each other, but there has been little demand by the water community in the past for guidance*
  - *There is an opportunity to do more on this, and we need an affirmative effort as opposed to waiting to be asked*
- *Identify the elements for decision tools. They should incorporate:*
  - *Low frequency extreme events*
  - *What do we do about what we see going on?*
  - *The paleoclimate record which gives a wider ranger of variability beyond the instrumental record.*
- A toolbox of climate tools (compendium) is being put together by the Global Water Research Coalition.
- *Tool box and guide needed for navigating climate change terminology, metrics, and protocols.*
- A glossary should be developed so that when decision-makers are being educated, uniform definitions are used.
- A primer on uncertainty should be developed and utilities should have access to it. The U.S. Climate Change Science Program and the IPCC have developed such a primer.
- Utilities should have a primer on uncertainty.
- Would like to see training materials made available to elected officials, especially regarding decision-making mechanisms.
- We need to develop a comprehensive summary on climate change for water infrastructure, including sustainability indicators for financial, social, and programmatic factors. MWCOG issued a climate change report in November 2008 that provides context and expectations.
- A taxonomy of approaches that are unique to different situations is needed. We need to outreach those that are lacking in knowledge.
- Not confident that local utilities have available standard guidelines or recommended strategies for addressing climate change. Such guidelines should include forming a team, developing a plan, doing an assessment, and implementing a plan. In addition, there needs to be a communication strategy and tool kit for utilities, including documents, case studies, terms, metrics, and protocols. Some information is out there from the Water Research Foundation and King County, but it needs to be centralized and packaged. This would give utilities a framework for thinking through the issues.
- Coastal states are beginning to realize the importance of looking at water, but interior states are lagging behind and are still concentrating almost exclusively on energy. There is a need for broad assessment screening tools to determine what certain climate change processes will do for water impacts in addition to all the other impacts.

Proceedings of the First National Expert and Stakeholder Workshop on
Water Infrastructure Sustainability and Adaptation to Climate Change

88

- Help is needed for smaller municipalities. A checklist might be too simple, but something to outline the process and considerations would be useful. The King County local government guidebook is a good place to start, but it lacks consensus. A centralized clearinghouse to provide resources for utilities is needed. There should be a local focus, and the clearinghouse would be helpful to more than just the utilities.

- What should we do with small regional utilities? There should be an easy to use "cookbook" that has information which is readily available and contains centralized data. We would then use the tools and data for better communication.

- There is a disconnect between universities and utilities. We need better application of research to the real world. There are opportunities for researchers and students to help with this.

- There needs to be a stronger link between the research and the utility operations communities, and we need to transfer knowledge of water management strategies to climate change.

- Is there a standard modeling technique used in developing a water footprint? The Peoria, Arizona planning commission presented their water footprint at the American Water Works Association (AWWA) Annual Conference and Exposition (ACE) in Atlanta.

- Do decision tools exist? Not in practice. [Delete because this observation is incorrect?]

Regional and Local Information and Tools

- A very simple web-based, location-specific tool should be developed that would allow a user to see the range of predictions of how precipitation and temperature might change over the next century. Projections could include a confidence interval of this tool's projections.

- There is a need for population, economic, and land cover forecasts at the state and local levels that are consistent with emissions scenario storylines. Information on state demography can be used to figure out where populations might be moving given certain climate change impacts.

- A detailed list of how climate change can affect a utility (e.g., storage tanks, water quality) needs to be developed.

- It would be helpful to have a tool and to run some forecasts through it, and then be able to enter observed data to test its

- Probability-based models that help utility managers with planning are needed (e.g., water quality or quantity models in a specific region).

Special Needs of Small Utilities

- *What are the real needs of small systems? How much is climate change going to impact them?*

- *Small systems may be more resilient, as they do not need 50-year investments, so there is not a lot to move*

- Where do these small engineers get their information now?

    - 1046 training is dying out, but a new way to educate them and provide new information has not been given to them

    - There are training centers and professional associations available for training.

- *Develop a checklist for small utilities derived from larger utilities' work*

Proceedings of the First National Expert and Stakeholder Workshop on
Water Infrastructure Sustainability and Adaptation to Climate Change

89

- *To address the gap between small-, mid-, and large-size utilities, develop guidance structured to be useable by thousands of smaller utilities*

- *Intermediate size is small in a climate change context*

- Help is needed for smaller municipalities. A checklist might be too simple, but something to outline the process and considerations would be useful. The King County local government guidebook is a good place to start, but it lacks consensus. A centralized clearinghouse to provide resources for utilities is needed. There should be a local focus, and the clearinghouse would be helpful to more than just the utilities.

- *CUPSS asset management tools, for example, was developed for small systems*

- *Homeland security was a similar issue for utilities.*

  - *First, vulnerabilities must be identified and risk assessment conducted.*

  - *Then, specific, tailored tools were developed for small utilities*

  - *They did not just downscale what the big utilities do, a simple, structured tool was created that met their needs.*

  - *A workforce was hired to work with small utilities to get it done. Would that make sense for climate change? Or hire a regional authority to do this?*

  - *But, homeland security issues were addressed by regulation, which meant it had to be done.*

- *For small systems, do research at a regional level and provide it as a consumable information*

- *Is climate change a way to re-stimulate small system sustainability?*

  - *Limited physical boundary*

  - *Cannot be physically interconnected*

  - *Bring sophisticated management, technical expertise, and financial support to disparate small systems; enforcement needs to get somewhat more aggressive to raise the stakes (typically enforcement gives them a by)*

- What should we do with small regional utilities? There should be an easy to use "cookbook" that has information which is readily available and contains centralized data. We would then use the tools and data for better communication.

- *Need a delivery mechanism for market penetration for small and medium utilities*

- *States are the mechanism for reaching small utilities*

- *Rural electrical cooperatives are buying small wastewater/drinking water systems to increase population to sell electricity to. They have money and technical ability to care for smaller systems. They are a player.*

- Big utilities are well staffed and aware of issues, but these issues need to be translated to medium- and small-sized utilities. Developing best practices would be the best option, particularly with engineering.

- *Medium to large utilities are aware of the issue, but we need to figure out how to engage the small to medium utilities. The last few years were focused on better management, and now climate change considerations need to be included while managing day-to-day responsibilities. We need to incorporate awareness of long-term factors.*

- *Big utilities tackle issues first and the challenge is to communicate the issues to smaller systems.*

Proceedings of the First National Expert and Stakeholder Workshop on Water Infrastructure Sustainability and Adaptation to Climate Change

90

- *Information flows down through associations and then picked up by medium size utilities.*

- *Information from associations has minimal impact on smaller facilities, though Rural Water representatives eventually pick up the information.*

- *Discussion of climate change and sustainable infrastructure are long-term projects and we need to keep talking. The National Rural Water Association needs to believe this first.*

- There is the issue of demographics, where people are moving to areas where there is no water. With the density in some urban areas, the infrastructure is not sustainable. Can small utilities be successful, or do we need to centralize them into larger utilities?

  - We have too much invested in wastewater treatment plants.

  - We have not had serious discussions about this; having these discussions may end up saving a lot of money.

Federal Role and Interactions with Federal Agencies

- There is a need for an ongoing stakeholder conversation that includes multiple feedback loops and helps guide research for the utility community. This present workshop is insufficient. There could be a standing committee, such as a standing advisory body or an earth systems science agency. There needs to be a much more forceful mandate.

- *Figure out how the multiple mission agencies can talk together with the multiple communities, not one at a time.*

- *Develop a steering committee or small group to keep the ball moving in the right direction, rather than the research community going off for three years without coordination.*

- *It needs to be a process, not a workshop, with a product at the end that builds.*

  - *This does not exist currently in the CCSP*

- *The process must not be too onerous. There are many different associations. Federal agencies need a way to listen to them all together. Coordination is needed on both sides.*

- *To turn the "fleet," the CCSP agencies should act in a coordinated fashion, and divide the labor reasonably.*

  - *This should be a coordinated effort*

  - *It should be transparent to the outside*

  - *It should be an ongoing, evolving, meaningful stakeholder process*

- *As CCSP evolves, we need to revisit this discussion to inform their work*

- *Coordination of agencies and transparency: as chairman of OSTP/SWAC, make sure this user community communicates with EPA*

- *NOAA, EPA, USGS, Bureau of Reclamation, and Corps of Engineers are interested in how they can serve the community of users. As an avenue of coordination and discussion, SWAC can help.*

Proceedings of the First National Expert and Stakeholder Workshop on Water Infrastructure Sustainability and Adaptation to Climate Change

91

- The federal government should consider a Bulletin 17B equivalent for climate change. This document was published to establish a set of procedures for estimating flood frequency and magnitude from coast to coast, with a purpose of establishing uniformity.

- Utilities need a coordinated and strategic federal science effort and this effort needs to be transparent. There needs to be a transparent plan for research that includes an organization chart detailing who is involved and what their roles are – this is currently indiscernible – and the people in the field need to understand what the federal government is doing.

- Interagency coordination is key, but such coordination might require a Congressional mandate, especially the creation of a national clearinghouse of information on climate change.

- There is a need for the federal government to provide an overview of information to top state agency people demonstrating why they should or should not be focused on the water side of climate change impacts.

- There is a need for a national climate service with a proper mission, subject to cooperative planning between all agencies. This service could provide a place to assemble and disseminate information, research, and tools and could be a repository for data. Much of the needed information that we have identified in this session (e.g., the need for collection and distribution) could be housed under a central "national climate service." This idea has been proposed in Congress.

- It appears that the rules in the approach of the Army Corps of Engineers to reservoir management are overly rigid and need more flexibility. We also need to balance the needs for recreational, drinking water, and other uses. Watershed management is the larger issue.

- We need to streamline processes for implementing alternate sources for water storage and supply. For instance, it is hard to control flooding with shallow reservoirs. Stricter zoning is also needed, in addition to funding for capital improvements.

- Utility managers (including federal managers such as the Bureau of Reclamation and the Army Corps of Engineers) need to increase their familiarity with risk-based decision tools rather than inflexible regulatory approaches. The risk-based tools need to be on timescales of days and decades.

- *Need regulatory flexibility - What are the regulatory implications associated with climate change?*

The Role of EPA

- *Target the right projects in order to move the "ship" in the direction of what is beneficial. Do not just research the old portfolio. Identify what EPA should do, on what it should partner, and where it should get out of the way.*

- The two prevailing paradigms from this workshop (i.e., the Lettenmaier – Behar – Estes-Smargiassi paradigm and the Brown – Waage paradigm) need to be reconciled with EPA's role, which is risk assessment, most of the time.

- EPA should be looking at the question, "How does climate change interface with the continuing implementation of Clean Water Act programs?"

- Regulators are accustomed to using the Clean Water Act regulations as their tool. The question is, therefore, "How does climate change interface with the continuing implementation of Clean Water Act programs?" For example, for monitoring programs, do we need guidelines for climate change? For water quality standards (e.g., in terms of temperature), do we revise the temperature standards?

Proceedings of the First National Expert and Stakeholder Workshop on
Water Infrastructure Sustainability and Adaptation to Climate Change

92

- *It will be difficult and complex for EPA to modify standards, and the discussion needs to start now.*

- Often utilities are working under 20-year consent decrees. There needs to be a paradigm shift for regulators that allows for shorter timeframes and the opportunity to reevaluate plans and designs periodically.

- *For negotiating consent decrees, permits, etc., what do EPA and state regulators need to be on same page with regard to climate change endpoints?*

- The traditional model of following consent degrees is inappropriate. There will need to be a shift from traditional infrastructure to sustainable infrastructure. It is possible that the current approach of following consent decrees will be counter-productive in the long-term.

- EPA needs to be funding a variety of impact studies and distribute the findings of these studies to the water community. All water quality parameters should be evaluated under different climate change scenarios.

- EPA should be supporting efforts to update precipitation frequency estimates.

- There is a need for better communication of standards of practice. A member of the climate modeling community seconded this sentiment, adding that there is also a need for sanctioned tools for water quality permits that puts the burden of uncertainty (liability) on the government.

- There is a need for assessments on changes to receiving waters as a result of climate change (e.g., acidification), and how that affects NPDES compliance for water quality. Regulatory changes will be needed, particularly in light of population changes.

- It would be helpful if EPA would develop a process for sanctioning tools for water quality and stream flow analysis (sanctioned in the sense that if a permittee agrees to implement the tools, the permittee will be deemed to be in compliance with the appropriate standards for a specified period of time). This would allow the government to specify that climate change parameters be included in analyses in some rational form. It would also allow individuals to be creative in developing solutions to problems and having a way to get them implemented and avoid the legal liability associated with tool uncertainty.

- EPA could conduct seminars and develop case studies of adaptation practices for utilities.

- EPA should be looking into developing climate protection levels for future time-slice projections. For example, EPA should be working towards being able to tell coastal cities what they can expect, and what they should protect against, in 2050, 2075, and 2100. EPA could say, for example, "If you are a coastal city you should plan for half a meter of sea level rise by 2050 as a conservative estimate."

- The drivers of the watershed approach (including EPA and other federal agencies) should formulate strategies by looking at not only intake but also at impacts and solutions. These drivers should consider institutional mandates and education to encourage action by utilities.

- *There are significant regulatory impacts of climate change (e.g., more violations), so will we need a new MCL? What will be the compliance issues?*

- *Concentrate on information to feed into the regulatory program (wetlands, water quality, aquifer systems)*

- EPA in the 1970s was facing the same issues we are today. There is no money available and there are no federal mandates. Economic incentives can give utilities the motivation to engage in energy efficiency and low impact development (LID) techniques.

- *ORD research is about aging infrastructure and the correlation between aging infrastructure and climate change adaptation. There will be opportunities over the coming decades.*
  - *The task for this generation is to set up the research problem correctly*
  - *Per Doug Owen, the real challenge is to get the direction of change and range of uncertainty. This is more important than timing.*
- *Water quality research is EPA's role. There is very little research in the literature on this. EPA needs to fund significant basic research on what is going to happen to water quality.*
- *Water/Energy nexus: EPA has an important role on wastewater treatment emissions of nitrous oxide and methane. Utilities can work on mitigation and adaptation with one lever. Use EPA's regulatory authority.*

Water Quality Implications

- There should be a characterization of a minimum base flow regime to support aquatic ecosystems. There is a nominal water supply, which is a limiting factor. It is important to establish a metric before putting it in place.
- There is a significant dearth of information available in through EPA that relates water quality impacts to climate change.
- There is a need for a standardization or sanctioning of tools for analyzing water quality and stream flow (sanctioned in the sense that if a permittee agrees to implement the tools, the permittee will be deemed to be in compliance with the appropriate standards for a specified period of time).
- Environmental groups are likely to sue if water regulators relax any regulations. We need to look closely at climate change vis-à-vis the practical implementation of supporting policies.
- There is very little literature on translating water quality parameters that utilities can use for operations under different climate change scenarios. Predicting total organic compounds in a river based on various climate change impacts, for example, would be a useful capability. Utility mangers would need such information to make decisions. We have some data that can be used (e.g., if climate shifted in this direction, here is what the impact would be), but the kind of information that is going to be necessary if water quality managers are going to be able to take action is not yet available.
- *Need analysis of wastewater and drinking water interactions (e.g., Cincinnati)*
  - *Water quality, drinking water, public health impacts*
  - *Stormwater/wastewater discharges*
- There are Global Change Research Program (GCRP) projects that are looking at several issues such as pH and sediments in hydrological modeling runs. There is a two-year timeframe before this information will be available. But there is an immediate need for hydrological water quality models that will offer users an ensemble of water quality parameters.
- There is concern at the local level over increasing invasive species prevalence as a result of climate change. How will these species spread over time and how should utilities handle them?
- Any place that has to meet total maximum daily load (TMDL) requirements has an understanding of certain standards. Otherwise, the understanding of these metrics is not as systematic.

Proceedings of the First National Expert and Stakeholder Workshop on
Water Infrastructure Sustainability and Adaptation to Climate Change

94

- Water providers fear that they will be blamed for any changes in the watershed, such as the rising temperature of a river. There is a need to better understand the watershed impact of utility decisions
- Research on treatment that is available and affordable is needed.

Water, Energy, and Greenhouse Gas Emissions

- Irrigation measures that are not maladaptive should be developed. If there is adaptation to mitigate greenhouse gases (GHGs), what will this exacerbate?
- For the most part, state agencies are not focused on the water side of climate change. Rather, they are focused on achieving GHG emissions reductions because the public is most concerned about this issue in the present.
- *The water footprint concept should be developed*
- There is a need for tools which can link adaptation and mitigation so that a utility can perform benefit-cost analyses which consider both types of options.
- *Identify best practices on how to fit adaptation and mitigation together.*
- There is a need for all involved stakeholder communities to reevaluate how they think about these issues. Should we be looking more closely at the water-energy nexus, for example? Regulators are eliminating electric utility disincentives for investing in energy efficiency and conservation measures by setting utilities' rates based on the amount of revenue they need to operate, regardless of ratepayer demand — essentially decoupling rates from demand. Should water utilities be looking at similar options? Also, looking at building codes as an example, are there ways to address water consumption at the building level that utilities can promote?
- Utilities should limit their carbon impact by minimizing the carbon used for the energy needed in their process. There is a need to optimize performance on energy conservation.
- *Utilities look to fund a number of solutions with each dollar, including future regulations, CO_2 footprint, etc. We need to foster this.*
- *Per EPRI, power is the #1 user of water, followed by irrigation then supply. Where can they partner, collaborate, or lead?*
- An analysis that shows the carbon-related impacts to compliance with water quality standards is needed. This should include both drinking water and wastewater. This will help analyze the "big picture" and promote discussions with regulators. This relates more to mitigation than adaptation, but some mitigation relates to adaptation in that mitigation affects options for adaptation.
- We need better models for post-discharge nitrification and denitrification to help determine the impacts.
- *Need protocols on how to measure emissions from wastewater treatment plants (e.g., nitrogen emissions)*
- *Need a water use metric for energy technologies*
- *Decentralized waste treatment and on-site treatment could be a strategy for adaptation*
    - *Need to verify IPCC assertions that they are a major source of methane*
    - *But package treatment plants were a problem, and would need vigorous operations and maintenance programs*

- *Identifying the linkages in soft infrastructure between drinking water and wastewater facilities at the local level (e.g., capital improvements planning). Climate change could be a translating mechanism about common interests, such as adaptation and mitigation challenges.*

Communication, Education, and Public Outreach

- Environmental groups need to be better represented in integrated resource planning processes. In California, a proposition was passed that motivated water agencies to include environmental groups in planning processes (e.g., by using grants). Climate change adaptation was a common ground for the two communities.

- Looking at the connection between climate change and water infrastructure leads to the question of financing. We have not yet connected to the public to explain the value of clean and stable water supplies. It is politically difficult to raise rates, yet customers prefer to go to the store to buy bottled water.

- Regional water master plans need to consider all affected groups and all stakeholders need to be involved in planning processes.

- There needs to be a two-phased approach based on answering the following questions: (1) Do I have a problem here? (2) If yes, what do I do about it? We need to determine who the target audience for outreach materials is – either the utilities themselves or their consulting engineers.

- With better outreach we can express investment as it relates to consumers. Financial information needs to be provided in a transparent manner, and the relationship between adaptation and capital investment needs to be made.

- We need to provide information on new technologies for water treatment, supply, and energy generation to local municipalities. Some technologies are mitigation driven, but they relate to adaptation.

- We need to better communicate all risks and risk management strategies, and improve communication with both management and stakeholders.

- More public communication in the form of a basic discussion on climate change is needed. For example, a local progressive water utility explained what they were doing, but did not bring up the subject of climate change. Very basic information is needed so we can relate climate change to other issues such as population growth.

- *Conduct market research at the local level on perceptions about adaptation*

  - *CCAP's Winkleman coined the term: NIMTO (not in my term of office)*

  - *Toronto, Chicago, New York City, and San Francisco have city-wide adaptation plans*

  - *Marketing is about selling vulnerabilities – health, transportation, planning, and land use*

  - *Cincinnati does a market research survey every two years and found that customers are willing to pay if they know where the money is going (i.e., to ensure a healthy, safe, plentiful water supply)*

- Climate change is not on the radar for sewer authorities. It is important to get information out, not in the form of textbooks, but in smaller mediums, such as state conferences, presentations, and small community training (particularly for design engineers).

- Utilities need to embrace education. Climate change should also be integrated with other issues, and all of these issues should be communicated to the public.

Proceedings of the First National Expert and Stakeholder Workshop on
Water Infrastructure Sustainability and Adaptation to Climate Change

96

- Communication needs to become more visual. When you engage the community, you can show them how to correctly value their water.

- The stakeholder process is very much affected by assumptions. The models often do not give an answer, or inform us about uncertainty. How do we get the process to the stakeholder level (i.e., going through the model with the stakeholder)? Stakeholders will engage and understand if they can vary inputs and learn how inputs change. We have observed this using the OASIS tool.

- *Promote the concept of boundary organizations to help translate climate science into useable information by working with local governments*

    - *For example, Arizona Water Research Institute and Colorado University Western Waters Assessment*

- *There is grassroots, municipal interest with greenhouse gas plans that are uniquely focused on the CO_2 footprint. This offers a point of entry for raising questions about adaptation.*

    - *What can a utility do to contribute to a community's mitigation goals? Or to contribute to adaptation?*

- *Maybe we need a FACA for structured, consistent guidance from non-federal employees.*

- *Organize communities of practice and gather stakeholder input.*

Other Recommended Ideas and Suggestions

- We need to learn more about the effects of the frost line and the roles utilities play during heat waves.

- *Some areas will not have water (e.g., groundwater). This is a fundamental water supply problem that cannot be fixed at a small scale*

Proceedings of the First National Expert and Stakeholder Workshop on
Water Infrastructure Sustainability and Adaptation to Climate Change

97

Appendix A   List of Workshop Participants

Vahid Alavian
Water Advisor
World Bank
valavian@worldbank.org
202-473-3602

Steve Allbee
Project Director, Gap Analysis,
Municipal Support Division
EPA
allbee.steve@epa.gov
202-564-0581

David Behar
San Francisco Public Utilities Commission
Staff Chair, Water Utility Climate Alliance
dbehar@sfwater.org
415-554-3221

Nancy Beller-Simms
Program Manager
NOAA Climate Program
nancy.beller-simms@noaa.gov
301-734-1205

Erika Berlinghof
Director of Congressional Relations
National Association of Water Companies
erika@nawc.com
202-833-8383

Rona Birnbaum
Chief, Climate Science and Impacts Branch
EPA OAR Climate Change Division
birnbaum.rona@epa.gov
202-343-9076

Pratim Biswas
Chair, Professor
Washington University
pratim.biswas@wustl.edu
314-935-5482

Geoff Bonnin
Chief, Hydrologic Science and Modeling
Branch
NOAA National Weather Services
geoffrey.bonnin@noaa.gov
301-713-0640 x103

Linda Boornazian
Director, Water Permits Division
EPA OW Office of Wastewater Management
boornazian.linda@epa.gov
202-564-0221

Levi Brekke
Professional Engineer
Bureau of Reclamation Technical Service
Center
LBREKKE@do.usbr.gov
303-445-2494

Barbara Brown
P.E., Strategic Leader
CDM Federal Programs Corporation
brownbs@cdm.com
617-452-6411

Casey Brown
Assistant Professor of Civil and Environmental
Engineering
University of Massachusetts
CBrown@ecs.umass.edu
413-577-2337

Erica Michaels Brown
Director, Regulatory Affairs
Association of Metropolitan Water Agencies
brown@amwa.net
202-331-2820

Ed Buchan
Environmental Coordinator
City of Raleigh, North Carolina
Edward.buchan@ci.raleigh.nc.us
919-857-4540

Proceedings of the First National Expert and Stakeholder Workshop on
Water Infrastructure Sustainability and Adaptation to Climate Change

98

Steven Buchberger
Professor
University of Cincinnati
Steven.Buchberger@uc.edu
513-556-3681

Brian Busiek
Senior Project Engineer
LimnoTech
bbusiek@limno.com
202-833-9140

Bob Cantilli
EPA OW Office of Science and Technology
cantilli.robert@epa.gov
202-566-1091

James Carleton
Lead Environmental Scientist
EPA OW Office of Science and Technology
Carleton.Jim@epa.gov
202 566-0445

Keith Cartnick
Director Water Quality and Compliance
United Water
Keith.Cartnick@UnitedWater.com
201-599-6031

Ni-Bin Chang
Professor
University of Central Florida
nchang@mail.ucf.edu
407-7547521

Jim Chelius
American Water Services

Robert Clark
Consultant
NRMRL-UC WRAP Team
rmclark@fuse.net
513-891-1641

Ann Codrington
EPA
codrington.ann@epa.gov
202-564-4688

Carol Collier
Executive Director
Delaware River Basin Commission
carol.collier@drbc.state.nj.us
609-883-9500, ext. 200

Elizabeth Corr
Associate Director, Drinking Water Protection
Division
EPA Office of Ground Water and Drinking
Water
corr.elizabeth@epa.gov
202-564-3750

John Cromwell
Environmental Economist
Stratus Consulting Inc.
jcromwell@stratusconsulting.com
202-741-1243

Andy Crossland
Sustainable Infrastructure Coordinator
EPA OW Office of Wastewater Management
crossland.andy@epa.gov
202-564-0574

Mark Crowell
FEMA
202-646-3432

Joshua Dickinson
Deputy Executive Director
WateReuse Foundation
jdickinson@watereuse.org
703-548-0880, ext. 104

Dennis Diemer
General Manager
East Bay Municipal Utility
dennisd@ebmud.com
510-287-0102

Cynthia Dougherty
Director
EPA Office of Ground Water and Drinking
Water
dougherty.cynthia@epa.gov
202-564-3750

*Proceedings of the First National Expert and Stakeholder Workshop on
Water Infrastructure Sustainability and Adaptation to Climate Change*

99

Jane Downing
Chief, Drinking Water Branch
EPA Region 1
downing.jane@epa.gov
617-918-1571

David Easterling
Chief, Scientific Services Division
NOAA National Climatic Data Center
david.easterling@noaa.gov
828-271-4675

Mikaela Engert
Planner
City of Keene, NH
mengert@ci.keene.nh.us
603-352-5474

Stephen Estes-Smargiassi
Director
Massachusetts Water Resources Authority
stephen.estes-smargiassi@mwra.state.ma.us
617-242-6000

Lauren Fillmore
Program Director
Water Environment Research Foundation
lfillmore@werf.org
703-684-2470, ext. 7153

Cynthia Finley
Director, Regulatory Affairs
National Association of Clean Water Agencies
cfinley@nacwa.org
202-296-9836

Mike Finn
EPA Office of Ground Water and Drinking
Water
finn.michael@epa.gov
202-564-5261

Josh Foster
Manager of Climate Adaptation
Center for Clean Air Policy
jfoster@ccap.org
202-408-9260, ext. 221

Mark Fulton
Global Head of Climate Change Investment
Research, Deutsche Asset Management
mark.fulton@db.com
212-454-7881

Alice Gilliland
Chief, Applied Modeling Research Branch
EPA ORD National Exposure Research
Laboratory
Gilliland.alice@epa.gov
919-541-0347

Peter Gleick
President
Pacific Institute
pgleick@pipeline.com
510-251-1600

James Goodrich
Acting Director
EPA ORD National Risk Management Research
Laboratory
goodrich.james@epa.gov
513-569-7605

Walter Grayman
Consultant; Principal, Owner
U. Cincinnati/Grayman & Assoc.
grayman@fuse.net
513-761-1722

Benjamin Grumbles
Assistant Administrator
EPA Office of Water
grumbles.benjamin@epa.gov
202-564-5700

Sally Gutierrez
Director
EPA ORD National Risk Management Research
Laboratory
gutierrez.sally@epa.gov
513-569-7683

Jim Hanlon
Office Director
EPA OW Office of Wastewater Management
hanlon.jim@epa.gov
202-564-0748

Proceedings of the First National Expert and Stakeholder Workshop on
Water Infrastructure Sustainability and Adaptation to Climate Change

100

Roy Haught
Water Quality Management Branch Chief
(Acting)
EPA ORD National Risk Management Research
Laboratory
haught.roy@epa.gov
513-569-7067

Steve Heare
Director, Drinking Water Protection Division
EPA Office of Ground Water and Drinking
Water
heare.steve@epa.gov
202-564-7992

Chuck Hennig
Research Coordinator
U.S. Bureau of Reclamation
chennig@do.usbr.gov
303-445-2134

John Henz
Atmospheric Group Leader
HDR Engineering
John.Henz@hdrinc.com
303-764-1582

Robert Hirsch
Research Hydrologist
U.S. Geological Survey
rhirsch@usgs.gov
703-648-5888

Rick Holmes
Director, Environmental Resources
Southern Nevada Water Authority
Richard.Holmes@snwa.com
702-862-3706

Heather Holsinger
Senior Fellow
Pew Center on Global Climate Change
holsingerh@pewclimate.org
703-516-0631

Raymond Jack
Director, Public Works, Town of Falmouth, MA
Town of Falmouth / Massachusetts Water
Works Association
capejack@capecod.net
508-457-2543

Laura Jacobson
Manager, System Operations, Planning
Division
Las Vegas Valley Water District
Laura.jacobsen@lvvwd.com
702-258-3186

Tom Johnson
Scientist
EPA ORD National Center for Environmental
Assessment
johnson.thomas@epa.gov
703-347-8618

Ernest Jolly
Energy Manager
DC Water and Sewer Authority
Ernest.Jolly@dcwasa.com
202-787-2370

Pamela Kenel
Associate Vice President, Global Water
Resources and Sustainable Planning Practice
Black & Veatch Corporation
kenelpp@bv.com
301-921-2885

Jim Kern
Environmental Engineer
EPA Region 3
kern.jim@epa.gov
215-814-5788

Joan Kersner
Manager
Drinking Water Planning and Peformance
Management Section
Seattle Public Utilities
joan.kersnar@seattle.gov
206-684-0839

Ephraim King
Office Director
EPA OW Office of Science and Technology
king.ephraim@epa.gov
202-566-0430

Proceedings of the First National Expert and Stakeholder Workshop on
Water Infrastructure Sustainability and Adaptation to Climate Change

101

Paul Kirshen
Professor
Tufts University
paul.kirshen@tufts.edu
617-627-5589

Chet Koblinsky
Director
NOAA Climate Program Office
chester.j.koblinsky@noaa.gov
301-734-1263

James LaGro
AAAS Science and Technology Policy Fellow
EPA National Center for Environmental
Assessment
lagro.james@epa.gov
703-347-8615

Cynthia Lane
Regulatory Engineer, Government Affairs Staff
American Water Works Association
clane@awwa.org
202-326-6122

Dennis Lettenmaier
Professor
University of Washington
dennisl@u.washington.edu
206-543-2532

Audrey Levine
National Program Director for Drinking Water
Research
EPA Office of Research and Development
levine.audrey@epa.gov
202-564-1070

Sylvana Li
Branch Chief, Rural Development and Natural
Resources
USDA
Sylvana.Li@usda.gov
202-690-2868

David Major
Professor
Columbia University
majorhart@earthlink.net
dcm29@columbia.edu
212-255-8329

Robert Marlay
Department of Energy
Robert.Marlay@hq.doe.gov
202-586-3949

Robyn McGuckin
Director of Strategic Planning
MWH Global Inc.
Robyn.mcguckin@mwhglobal.com
303-533-1976

Linda Mearns
Senior Scientist
National Center for Atmospheric Research
lindam@ucar.edu
303-497-8124

G. Tracy Mehan
Principal, Drinking Water and Water Quality
Group
Cadmus, Inc.
gmehan@cadmusgroup.com
703-247-6106

Karen Metchis
EPA Office of Water Transition Coordinator
EPA OW Office of Wastewater Management
metchis.karen@epa.gov
202-564-0734

Jami Montgomery
AAA Fellow
EPA ORD Office of Science Policy
montgomery.jami@epa.gov
202-564-0693

Dean Moss
General Manager
Beaufort-Jasper Water and Sewer Authority
DeanM@bjwsa.org
843-987-9210

Peter Mulvaney
Assistant Commissioner
Chicago Department of Water Management
peter.mulvaney@cityofchicago.org
312-744-3436

Proceedings of the First National Expert and Stakeholder Workshop on
Water Infrastructure Sustainability and Adaptation to Climate Change

102

Armin Munevar
Professional Engineer
CH2M Hill
armin.munevar@ch2m.com
619-687-0110

Daniel Murray
Senior Advisor for Water Quality
EPA Office of Research and Development
murray.dan@epa.gov
513-569-7522

Mike Muse
EPA Office of Ground Water and Drinking
Water
muse.mike@epa.gov
202-564-3892
Jill Neal
Environmental Engineer
EPA ORD National Risk Management Research
Laboratory
neal.jill@epa.gov
513-569-7277

Chuck Noss
National Program Director for Water Quality
EPA Office of Research and Development
noss.charles@epa.gov
919-541-1322

Rachel Novak
ORISE Intern
EPA OW Office of Science and Technology
Novak.Rachael@epa.gov
202-566-2385

Rolf Olsen
Senior Scientist
USACE Institute for Water Resources
j.rolf.olsen@usace.army.mil
703-428-6314

Doug Owen
Vice President and Chief Technology Officer
Malcolm Pirnie, Inc.
DOwen@PIRNIE.COM
914-641-2700

Kenan Ozekin
Senior Project Manager, Research
Management
Water Research Foundation
kozekin@awwarf.org
303-734-3464

Andrew Parker
Director, Water Resources Group
Tetra Tech Inc.
andrew.parker@tetratech.com
703-385-6000

William Perkins
EPA Office of Air and Radiation
perkins.william@epa.gov
202-343-9460

Jeff Peterson
Senior Policy Advisor
EPA Office of Water
peterson.jeff@epa.gov
202-564-5771

Jan Pickerel
Environmental Protection Specialist
EPA OW Office of Wastewater Management
Pickrel.Jan@epamail.epa.gov
202-564-7904

Tony Quintanilla
Assistant Director of Maintenance and
Operations
Metropolitan Water Reclamation District of
Greater Chicago
antonio.quintanilla@mwrd.org
207-247-8024

David Rager
Director
Greater Cincinnati Water Works
david.rager@gcww.cincinnati-oh.gov
513-591-7700

Rob Renner
Executive Director
Water Research Foundation
rbrenner@awwarf.org
303-347-6150

*Proceedings of the First National Expert and Stakeholder Workshop on
Water Infrastructure Sustainability and Adaptation to Climate Change*

103

Matt Ries
Managing Director of Technical and
Educational Services
Water Environment Federation
mries@wef.org
703-684-2400, ext. 7255

Frank Roth
Senior Policy Manager
Albuquerque Bernalillo County Water Utility
Authority
froth@abcwua.org
505-768-2511

Kellie Rotunno
Director of Engineering & Construction
Northeast Ohio Sewer District
216-881-6600

Mary Ann Rozum
Program Leader, Conservation and
Environment
USDA CSREES
MROZUM@CSREES.USDA.GOV
202-401-4533

Suzanne Rudzinski
Deputy Office Director
EPA OW Office of Science and Technology
rudzinski.suzanne@epa.gov
202-566-0430

Paul Rush
Deputy Commissioner, Bureau of Water
Supply
New York City Department of Environmental
Protection
PRush@dep.nyc.gov
845-340-7514

Carol Russell
Climate Change and Water Coordinator
EPA Region 8
russell.carol@epa.gov
303-312-6310

Greg Sayles
Associate Director
EPA ORD Homeland Security Research Center
sayles.gregory@epa.gov
513-569-7607

Joel Scheraga
National Program Director
Global Change Research Program
EPA Office of Research and Development
scheraga.joel@epa.gov
202-564-3385

Peter Schultz
Director
U.S. Climate Change Science Program
pschultz@usgcrp.gov
202-223-6262

Chi Ho Sham
Vice President
The Cadmus Group, Inc.
CSham@cadmusgroup.com
617-673-7156

Mike Shapiro
Deputy Assistant Administrator
EPA Office of Water
shapiro.mike@epa.gov
202-564-5700

Daniel Sheer
President
HydroLogics, Inc.
dsheer@hydrologics.net
410.715.0555

Mark Simpson
Water Division Manager
Manatee County Utilities Department
mark.simpson@mymanatee.org
941-792-8811 Ext. 5258

Joel Smith
Vice President
Stratus Consulting Inc.
jsmith@stratusconsulting.com
303-381-8218

Thomas Speth
Director, Water Supply Water Resources
Division
EPA ORD National Risk Management Research
Laboratory
speth.thomas@epa.gov
513-569-7208

Proceedings of the First National Expert and Stakeholder Workshop on
Water Infrastructure Sustainability and Adaptation to Climate Change

104

Neil Stiber
Environmental Scientist
EPA Office of the Science Advisor
stiber.neil@epa.gov
202-564-1573

Nancy Stoner
Clean Water Project Director
NRDC Clean Water Network
nstoner@nrdc.org
202-289-2394

Susan Sullivan
Deputy Executive Director
New England Interstate Water Pollution
Control Commission
ssullivan@neiwpcc.org
978-323-7929

Jim Taft
Executive Director
Association of Drinking Water Administrators
jtaft@asdwa.org
703-812-9507

Claudio Ternieden
Assistant Director of Research
Water Environment Research Foundation
cternieden@werf.org
703-684-2470, Ext. 7907

Ed Thomas
Engineer
National Rural Water Association
ruralwater@gmail.com
443-739-1358

Ed Torres
Director of Technical Services
Orange County Sanitation District
etorres@ocsd.com
714-593-7080

Brad Udall
Director
Western Water Assessment
bradley.udall@colorado.edu
303-497-4573

Betsy Valente
EPA OW Office of Ground Water and Drinking
Water
valente.betsy@epa.gov
202-564-9895

Paula VanHaagen
Manager, Grants & Strategic Planning Unit
EPA Region 10
vanhaagen.paula@epa.gov
206-553-6977

Alan Vicory
Execuitve Director and Chief Engineer
Ohio River Valley Water Sanitation
Commission
avicory@orsanco.org
513-231-7719

Marc Waage
Manager of Water Resource Planning
Denver Water / Water Utility Climate Alliance
Marc.Waage@DenverWater.org
303-628-6572

Chris Weaver
Physical Scientist
EPA ORD National Center for Environmental
Assessment
weaver.chris@epa.gov
703-347-8621

Laura Wharton
Supervisor, Comprehensive Planning and
Asset Management
Program Development
King County (Washington) Department of
Natural Resources and Parks
Laura.Wharton@kingcounty.gov
206-684-1238

Pai-Yei Whung
EPA Chief Scientist
EPA Office of the Science Advisor
Whung.Pai-Yei@epa.gov
202-564-0789

Proceedings of the First National Expert and Stakeholder Workshop on
Water Infrastructure Sustainability and Adaptation to Climate Change

105

Allison Wiedeman
Chief, Rural Branch, Permits Division
EPA OW Office of Wastewater Management
wiedeman.allison@epa.gov
202-564-0991

Jeff Yang
Scientist
EPA ORD National Risk Management Research
Laboratory
yang.jeff@epa.gov
513-569-7655

David Yates
Project Scientist
National Center for Atmospheric Research
yates@ucar.edu
303-497-8394

Doug Yoder
Deputy Director
Miami-Dade Water and Sewer Department
yoderd@miamidade.gov
786-552-8979

Phil Zahreddine
Chief, Municipal Technology Branch
EPA OW Office of Wastewater Management
Zahreddine.Phil@epa.gov
202-564-0587

Proceedings of the First National Expert and Stakeholder Workshop on
Water Infrastructure Sustainability and Adaptation to Climate Change

106

Appendix B   Workshop Agenda

Tuesday, January 6, 2009	
7:30	**Registration and Continental Breakfast**
Welcome and Introduction *Location: Ballroom A*	
8:00	**Welcome and Administrative Remarks** *Jim Hanlon, Director, EPA Office of Wastewater Management* *Cynthia Dougherty, Director, EPA Office of Ground Water and Drinking Water*
8:10	**Welcome, Workshop Objectives, and ORD's Commitment to Climate Change** *Sally Gutierrez, Director, EPA ORD National Risk Management Research Laboratory*
8:20	**The EPA National Water Program's Support for Climate Change Adaptation** *Benjamin Grumbles, Assistant Administrator, EPA Office of Water*
Plenary Sessions *Location: Ballroom A*	
I – Challenges and Opportunities in Adapting to Climate Change *Moderator: Dr. Pai-Yei Whung, Chief Scientist, EPA Office of the Science Advisor*	
8:30	**Adaptation Challenges to the Nation and the Science Community** *Dr. Peter Gleick, Pacific Institute*
8:45	**Perspectives from Utilities** *David Behar, San Francisco Public Utilities Commission / Staff Chair, Water Utility Climate Alliance*
9:00	**When R&D Meets the Real World: The Challenges and Opportunities of Integrating Water Resource Management for a Changing Climate** *Dr. James Goodrich, EPA ORD National Risk Management Research Laboratory*
9:15	Questions & Answers and Discussion
II – Applying Climate Science to Water Infrastructure Planning *Moderator: Jim Taft, Executive Director, Association of State Drinking Water Administrators*	
9:30	**Where the Research Meets the Road: Climate Science, Uncertainties, and Knowledge Gaps** *Dr. Dennis Lettenmaier, University of Washington*
9:45	**Information Needed for Infrastructure Adaptation Planning** *Stephen Estes-Smargiassi, Massachusetts Water Resources Authority*
10:00	**Accommodating Design Uncertainties: Past Practices and Future Needs** *Doug Owen, P.E., Malcolm Pirnie Inc.*
10:15	**Holistic ORD Research to Ensure Water and Energy Efficiency through Drinking Water System Sustainability** *Dr. Audrey Levine, P.E., EPA Office of Research and Development*
10:30	Questions & Answers and Discussion
10:45	*Break*
III – R&D for Water Infrastructure Adaptation *Moderator: Carol Collier, Executive Director, Delaware River Basin Commission*	
11:00	**EPA's Global Climate Change Science Program and Water Infrastructure Adaptation Research** *Dr. Joel Scheraga, EPA Office of Research and Development*
11:15	**AWWARF Research Strategy for Climate Change Adaptation**

Proceedings of the First National Expert and Stakeholder Workshop on
Water Infrastructure Sustainability and Adaptation to Climate Change

107

	David Rager, Greater Cincinnati Water Works
11:30	**WERF's Climate Change Research Programs** *Claudio Ternieden, Water Environment Research Foundation*
11:45	**Incorporating Climatic Uncertainties into Water Planning** *Marc Waage, Denver Water / Water Utility Climate Alliance*
12:00	Questions & Answers and Discussion
12:30	***Lunch*** *(on your own at either the hotel restaurant or a nearby establishment)*

	Concurrent Session Track A ***Climate Change Impacts on Hydrology*** ***and Water Resource Management*** *Location: Ballroom B*	**Concurrent Session Track B** ***Adaptive Management and Engineering:*** ***Information and Tools*** *Location: Ballroom C*
1:30	**A.1. Projecting Hydroclimatic Changes - Part I: Downscaling** *Moderator: Linda Mearns, University Corporation for Atmospheric Research*	**B.1. National Infrastructure Condition Assessment and Adaptability** *Moderator: Dr. Neil Stiber, EPA Office of the Science Advisor*
1:30	**Downscaling or Decision-scaling? An Overview of Downscaling** *Dr. Casey Brown, University of Massachusetts*	**Rehabilitation, Replacement, and Redesign of the Nation's Water and Wastewater Infrastructure as a Valuable Adaptation Opportunity** *Dan Murray, P.E., EPA National Risk Management Research Laboratory*
1:45	**Dynamic Downscaling Efforts at EPA: Regional Linkages to NOAA and NASA Global Scale Models** *Dr. Alice Gilliland, EPA National Exposure Research Laboratory*	**Flood Control and Surface Water Management Infrastructure in the Age of Climate Change** *Dr. Rolf Olsen, U.S. Army Corps of Engineers Institute for Water Resources*
2:00	**Web-Archive of Statistically Downscaled Climate Projections for the Contiguous United States** *Levi Brekke, P.E., Bureau of Reclamation Technical Services Center*	**Climate Change Readiness Assessment and Planning for the Nation's Drinking Water and Wastewater Utilities** *Dr. Steven Buchberger, P.E., NRMRL-UC WRAP Team*
2:15	**The North American Regional Climate Change Assessment Program: A Brief Overview** *Linda Mearns, University Corporation for Atmospheric Research*	**Assessing the Impacts of Climate Change on Drinking Water Treatment** *Dr. Robert Clark, P.E., NRMRL-UC WRAP Team*
2:30	Questions & Answers and Discussion	Questions & Answers and Discussion
3:00	***Break***	
3:30	**A.2. Projecting Hydroclimatic Changes - Part II: Local Applications of Downscaling** *Moderator: David Easterling, NOAA National Climatic Data Center*	**B.2. Progressive Adaptation: Planning and Engineering for Sustainability** *Moderator: Steve Allbee, EPA Office of Wastewater Management*
3:30	**Regional Modeling for the Pacific Northwest** *Dr. Dennis Lettenmaier, University of Washington*	**Overview: Integrating Climate Adaptation into Lifecycle Costing and Planning** *Steve Allbee, EPA Office of Wastewater Management*

Proceedings of the First National Expert and Stakeholder Workshop on
Water Infrastructure Sustainability and Adaptation to Climate Change

108

3:50	**Hydroclimatic Modeling for Water Resources Planning in the City of New York** *Dr. David Major, Columbia University*	**Adaptation of Water Infrastructure Investments to Changing Demands and Climate Variability: A Systems Approach** *Dr. Vahid Alavian, World Bank*
4:10	**Predictive Capacity in the Colorado River Basin** *Brad Udall, University of Colorado – NOAA Western Water Assessment*	**A Review of Quantitative Methods for Evaluating Impacts of Climate Change on Urban Water Infrastructure** *Dr. Walter Grayman, P.E., NRMRL-UC WRAP Team*
4:30	Questions & Answers and Discussion (60 minute session)	**Water Use and Re-Use in Energy Technologies in a Carbon-Constrained World** *Dr. Pratim Biswas, P.E., Washington University in St. Louis*
		Questions & Answers and Discussion (40 minute session)
5:30	*Adjourn*	

	Concurrent Session Track A *Climate Change Impacts on Hydrology and Water Resource Management* Location: Ballroom B	Concurrent Session Track B *Adaptive Management and Engineering: Information and Tools* Location: Ballroom C
8:00	*Registration and Continental Breakfast*	
8:30	**A.3. Evaluating Hydroclimatic Change for Water Infrastructure Adaptation - Part I** *Moderator: Dr. Daniel Sheer, HydroLogics, Inc.*	**B.3. Adaptation Practices and Tools - Part I** *Moderator: Josh Foster, Center for Clean Air Policy*
8:30	**Hydrology and Climate Change: What Do We Actually Know?** *Dr. Robert Hirsch, U.S. Geological Survey*	**Alternative Water Supply and Drinking Water System Operations: Preparation for Climate Change Adaptation in East Bay MUD** *Dennis Diemer, East Bay MUD, CA*
8:50	**Precipitation Frequency Atlas of the United States: Update and Issues** *Geoffrey Bonnin, NOAA National Weather Service*	**Stormwater Management and Extreme Precipitation: Protecting Surface Water and Source Water Quality in Ohio River Watersheds** *Alan Vicory, Ohio River Valley Water Sanitation Commission*
9:10	**National Hydroclimatic Change and Infrastructure Assessment: Region-Specific Adaptation Factors** *Dr. Y. Jeffrey Yang, P.E., EPA National Risk Management Research Laboratory*	**Case Study: Risk and Management Analysis for Progressive Adaptation of Water Supply in Metro Boston** *Dr. Paul Kirshen, Tufts University*
9:30	Questions & Answers and Discussion	Questions & Answers and Discussion

Proceedings of the First National Expert and Stakeholder Workshop on Water Infrastructure Sustainability and Adaptation to Climate Change

109

Wednesday, January 7, 2009		
10:00	*Break*	
10:30	**A.4. Evaluating Hydroclimatic Changes for Water Infrastructure Adaptation - Part II** *Moderator: Dave Behar, San Francisco Public Utilities Commission / Water Utility Climate Alliance*	**B.4. Adaptation Practices and Tools - Part II** *Moderator: Mikaela Engert, City of Keene, New Hampshire*
10:30	**Climate Vulnerability Assessments** *David Yates, National Center for Atmospheric Research* (20 minute presentation)	**Integrated Water Management for Sustainable Water Supply in SW Florida under Global Changes: Water Reuse and Energy Considerations** *Mark Simpson, Manatee County Manatee County Utilities Department* (15 minute presentation)
	Strategies for Assessing Impacts and Adapting to Climate Change for Wastewater Utilities *Laura Wharton, King County Department of Natural Resources and Parks* (20 minute presentation)	**EPA Water Resource Adaptation Program (WRAP) R&D Activities on Adaptation Methods and Techniques** *Roy Haught, EPA National Risk Management Research Laboratory* (15 minute presentation)
		BASINS CAT, WEPPCAT, and ICLUS: Modeling Tools for Assessing Watershed Sensitivity to Climate and Land Use Change *Dr. Tom Johnson, EPA National Center for Environmental Assessment* (15 minute presentation)
	Implicit Climate Change Adaption: Modifying System Operations for Turbidity Control *Paul Rush, NYC Bureau of Water Supply, and Dr. Daniel Sheer, HydroLogics, Inc.* (20 minute presentation)	**Metropolitan Water Availability Forecasting Methods and Applications in South Florida** *Dr. Ni-Bin Chang, P.E., University of Central Florida, NRMRL-UC WRAP Team* (15 minute presentation)
11:30	Questions & Answers and Discussion	Questions & Answers and Discussion
12:15	*Lunch (on your own at either the hotel restaurant or a nearby establishment)*	
Break-Out Discussions		
1:15	**Instructions for Breakout Sessions** *Moderator: John Cromwell, Stratus Consulting* *Location: Ballroom A*	
1:30	**Break-Out Discussions** Participants will divide into four separate discussion groups, two to focus on Track A issues and two on Track B issues. Each group selects a representative to summarize discussions and report out to the final plenary.	

Proceedings of the First National Expert and Stakeholder Workshop on Water Infrastructure Sustainability and Adaptation to Climate Change

110

Wednesday, January 7, 2009		
	Track A Break-Out Discussions *Moderators:* • *Joel Smith, Stratus Consulting* • *Karen Metchis, EPA OW* *Locations: Crystal III & Crystal IV*	Track B Break-Out Discussions *Moderators:* • *John Cromwell, Stratus Consulting, and Jeff Yang, EPA ORD* • *Jim Goodrich, EPA ORD, and Elizabeth Corr, EPA OW* *Locations: Private Dining Room (1st Floor) & Crystal II*
Plenary Session: Discussion and Concluding Remarks *Location: Ballroom A*		
4:00	**Report Out on Break Out Discussions** *Break-out Discussion Moderators*	
4:20	**Concluding Remarks** *Dr. Pai-Yei Whung, Chief Scientist, EPA Office of the Science Advisor*	
4:30	**Concluding Remarks** *Benjamin Grumbles, Assistant Administrator, EPA Office of Water*	
4:45	**Adjourn**	

Proceedings of the First National Expert and Stakeholder Workshop on
Water Infrastructure Sustainability and Adaptation to Climate Change

111

Appendix C   Biographies of Workshop Speakers and Moderators

Dr. Vahid Alavian, World Bank

Vahid Alavian serves as the Water Advisor at the World Bank where he is responsible for advising on complex investments in the water sector, leading dialogue on water resources management with major World Bank clients, and for helping implement the Bank's water resources strategy. Dr. Alavian has more than 30 years of experience in the water sector through work with international financial institutions, governments, donor agencies, private sector, and academia. He has led water-related projects and programs including water resources management, water supply and sanitation, hydropower, dam safety, and water quality and environmental compliance in a number of developing countries. He currently leads an analytical and advisory study on the potential impact of hydrologic variability and climate change on the Bank's water investments and adaptation measures to make these investments climate-smart. Prior to joining the World Bank in 2000, he served as the Senior Water Advisor at the United States Agency for International Development (USAID) Global Environment Center, where he helped advance sustainable management of freshwater water and coastal resources. As a Senior Specialist at the Tennessee Valley Authority, he led some of the agency's pioneering work on integrating water, energy, and environmental management for sustainable development and growth. He is a Fulbright Scholar and has held faculty positions at the University of Illinois, the University of Tennessee, and the University of Zambia.

Steve Allbee, EPA Office of Wastewater Management

Steve Allbee has been with EPA for 28 years, during which time he has held several senior positions in the Office of Water. Currently, as Project Director of the Gap Analysis, Mr. Allbee is the principal author of The Clean Water and Drinking Water Infrastructure Gap Analysis. The "Gap Analysis" is a comprehensive national-level assessment, published by EPA in September 2002, and is often cited as a primary source document in articulating the challenges ahead for America's water and wastewater systems. Of late, the central point of his work is on promoting advanced asset management approaches as a pathway toward sustainable water and wastewater services for the 21st Century. During his tenure at EPA, he has served as the Director of the Planning and Analysis Division, the Acting Director of the Municipal Construction Division, Chief of the Municipal Assistance Branch, Expert Advisor to the Border Environment Cooperation Commission and the North American Development Bank as well as undertaking several headquarters' staff assignments. Prior to joining EPA, he managed the planning of a large regional wastewater system with a service population of approximately 2 million people. He had national leadership responsibility for establishing the innovative State Revolving Fund (SRF) Program as a means to provide Federal financial assistance to wastewater infrastructure projects. He has also had the distinction of developing important special infrastructure assistance programs targeted to underserved and economically disadvantaged communities; including Mexico border communities, Tribes and Alaskan Native Villages. In addition, he has managed a broad network of technical assistance services that provide operations, maintenance and related support to small communities. He frequently provides technical assistance to international organizations on issues concerning water and wastewater organizations, project development, finance and management. He received a Masters of Public Administration from Harvard University, a Masters in Urban and Regional Planning from Mankato State University and a Bachelors of Arts in Political Science from Winona State University.

David Behar, San Francisco Public Utilities Commission / Water Utility Climate Alliance

David Behar's career spans twenty years in environmental advocacy, policy analysis, and water utility management. Mr. Behar currently serves as Deputy to the Assistant General Manager, Water Enterprise, at the San Francisco Public Utilities Commission. The SFPUC is the sixth largest municipal water provider in the U.S. and manages water and power facilities and operations at Hetch Hetchy, the regional system that delivers water 160 miles to 2.4 million Bay Area residents, and water,

Proceedings of the First National Expert and Stakeholder Workshop on
Water Infrastructure Sustainability and Adaptation to Climate Change

112

wastewater, and stormwater facilities in San Francisco. At the SFPUC, he manages business planning, local resource management strategies, and climate change adaptation planning. He led development of the SFPUC-sponsored Water Utility Climate Change Summit held in San Francisco in early 2007 and currently serves as staff chair of the Water Utility Climate Alliance (WUCA). Established formally in early 2008, WUCA is a coalition of eight water utilities dedicated to providing leadership and collaboration on climate change issues affecting drinking water utilities by improving research, developing adaptation strategies and creating mitigation approaches to reduce greenhouse gas emissions. WUCA is chaired by SFPUC General Manager Ed Harrington and includes Denver Water, the Metropolitan Water District of Southern California, New York City Department of Environmental Protection, Portland Water Bureau, San Diego County Water Authority, Seattle Public Utilities and the Southern Nevada Water Authority. Prior to joining the SFPUC, he was an environmental policy consultant whose clients included the Natural Resources Defense Council and the Pacific Rivers Council. From 1991-97 he served as Executive Director of The Bay Institute of San Francisco, and from 1989-91 he served on the staff of U.S. Senator Alan Cranston (D-CA). In November 2006 he was elected to the Board of Directors of the Marin Municipal Water District, a 200,000-customer water district just north of San Francisco in Marin County, where he lives with his two children. He has a bachelor's degree in politics from the University of California, Santa Cruz.

Dr. Pratim Biswas, P.E., Washington University in St. Louis
Pratim Biswas received his B.Tech. degree from the Indian Institute of Technology, Bombay in Mechanical Engineering in 1980; his M.S. degree from the University of California, Los Angeles in 1981; and his doctoral degree from the California Institute of Technology in 1985. After receiving his doctoral degree, he joined the University of Cincinnati as an Assistant Professor in the Environmental Engineering Science Division in 1985. He was promoted to Associate Professor in 1989, and became Full Professor in 1993. He also served as the Director of the Environmental Engineering Science Division at the University of Cincinnati for four years. In the interim, he spent a year's sabbatical at the National Institute of Standards and Technology in their Chemical Sciences and Technology Division in 1994. He joined Washington University in St. Louis in August 2000 as the inaugural Stifel and Quinette Jens Professor and Director of the Environmental Engineering Science Program. In 2006, he became the Chair of the newly created Department of Energy, Environmental and Chemical Engineering at Washington University in St. Louis. He has won several Teaching and Research Awards: was the recipient of the 1991 Kenneth Whitby Award given for outstanding contributions by the American Association for Aerosol Research; and the Neil Wandmacher Teaching Award of the College of Engineering in 1994. He was elected as a Fellow of the Academy of Science, St. Louis in 2003. Dr. Biswas is a member of the Steering Committee of the McDonnell International Scholars Academy, and the Ambassador to the Indian Institute of Technology, Bombay.

His research and educational interests are in aerosol science and technology, nanoparticle technology, energy and environmental nanotechnology, air quality and pollution control and the thermal sciences. He has published more than 170 refereed journal papers and presented more than 100 invited talks all across the globe.

Geoffrey Bonnin, NOAA National Weather Service
Geoff Bonnin is Chief of the Hydrologic Science and Modeling Branch of NOAA, National Weather Service, Office of Hydrologic Development. Mr. Bonnin manages science and technique development for flood and stream flow forecasting, and water resources services provided by the National Weather Service. The work of the group includes development and maintenance of U.S. precipitation frequency estimates. He initiated the development of NOAA Atlas 14 and was lead author for the first three volumes. He graduated with a B.E. (Civil) from the University of Queensland, Australia and a M.S. (Engineering Management) from the University of Kansas. He is a Chartered Member of the Institution of Engineers Australia and a member of the American Society of Civil Engineers. He

Proceedings of the First National Expert and Stakeholder Workshop on
Water Infrastructure Sustainability and Adaptation to Climate Change

113

has extensive experience in flood forecasting and flood forecast systems development with the U.S. National Weather Service and the Australian Bureau of Meteorology. He also has extensive experience in software engineering and systems integration in private industry. His primary areas of expertise are in data management as the integrating component of end-to-end systems, the science and practice of real time hydrologic forecasting, estimation of extreme precipitation climatologies, and the management of hydrologic enterprises. He is one of the developers, and the primary implementer, of Standard Hydrometeorological Exchange Format (SHEF).

Dr. Levi Brekke, P.E., Bureau of Reclamation Technical Services Center
Levi Brekke has been working with Reclamation since 2003 and currently works at Reclamation's Technical Service Center in Denver. Dr. Brekke's work focuses on reservoir systems analysis, technical team coordination, and conducting research on climate information applications. His education includes a B.S.E. in Civil Engineering (The University of Iowa), a M.S. in Environmental Science and Engineering (Stanford University), and a Ph.D. in Water Resources Engineering (University of California Berkeley). His work experience also includes consulting in the areas of wastewater and water treatment engineering where his efforts focused on capital improvements planning.

Dr. Casey Brown, P.E., University of Massachusetts
Casey Brown is assistant professor of Civil & Environmental Engineering at UMass, Amherst and Adjunct Associate Research Scientist at the International Research Institute for Climate and Society of the Earth Institute at Columbia University. Dr. Brown specializes in climate risk management for the water sector and sustainable management of water resources. His research focuses on increasing the resilience of water systems to climate variability and change through the use of advanced climate science and hydrologic forecasting, in combination with innovative water resources management techniques and economic mechanisms, including index insurance. Another area of interest is the role of climate variability, infrastructure and water management in poverty reduction and economic development. He is PI and co-PI for several projects in the U.S. and abroad funded by NOAA, the World Bank and other agencies and is a 2007 recipient of the Presidential Early Career Award for Scientists and Engineers. He is Associate Editor of the ASCE Journal of Water Resources Planning and Management and has published in Water Resources Research, Natural Resources Forum, International Journal of Climatology and the ASCE Journal of Water Resources Planning and Management, where a 2006 paper won the award for Best Policy-Oriented paper. He obtained his PhD in environmental engineering science as a National Science Foundation Fellow at Harvard University in 2004. He is a licensed professional engineer in the state of Colorado and a former U.S. Air Force officer.

Dr. Steven Buchberger, P.E., NRMRL-UC WRAP Team
Steven Buchberger is a Professor of Civil and Environmental Engineering at the University of Cincinnati where he has served on the faculty for 21 years, including a recent stint as interim department head. He earned his BS from the University of Wisconsin at Madison, his MS at Colorado State University and his PhD from the University of Texas at Austin – all in Civil and Environmental Engineering.

Dr. Buchberger is a registered professional engineer in the State of Colorado and a member of ASCE, AGU, AWWA, ASEE. He served as Associate Editor of the Journal *of Water Resources Planning and Management* for ten years and is a founding member of the organizing committee for the international symposium on Water Distribution Systems Analysis, now entering its 11th year with the ASCE EWRI conference circuit. For the past ten years Dr. Buchberger has managed a training grant providing opportunities for over 100 university students in engineering and science to pursue research at the EPA national laboratory in Cincinnati.

Proceedings of the First National Expert and Stakeholder Workshop on
Water Infrastructure Sustainability and Adaptation to Climate Change

114

Dr. Buchberger has over 115 publications in journals and proceedings, including chapters for a McGraw Hill *Handbook on Water Supply Systems Security*. He has won Young Investigator Awards from the National Science Foundation and from the Department of Energy and he received the Neil Wandmacher Senior Faculty Teaching Award from the College of Engineering. Dr. Buchberger has advised 46 graduate students; eleven have been recognized with best paper and/or best poster awards from ASCE, AGU, and AWWA.

Dr. Buchberger's teaching interests include surface water hydrology and reliability analysis in engineering design. His research interests are broad, but a favorite focus is mathematical modeling of water demands and water quality in municipal distribution systems. Here, Dr. Buchberger and his students demonstrated that many water demand patterns behave like a non-stationary Poisson rectangular pulse process. This led to the development PRPsym, the first computer code capable of generating high resolution stochastic water demands for urban network simulation. The PRPsym code has proven to be a valuable tool in studies of infrastructure vulnerability for homeland security.

Dr. Ni-Bin Chang, P.E., University of Central Florida, NRMRL-UC WRAP Team
Ni-Bin Chang was educated at the National Chiao-Tung University (NCTU) in Taiwan where he received his bachelor degree in Civil Engineering in 1983. Later on, Dr. Chang came to the United States in 1987 and received his Master's and Ph.D. degrees in the field of Environmental Systems Engineering at Cornell University in 1989 and 1991, respectively in the US. At present, he is a professor with Civil, Environmental, and Construction Engineering Department, University of Central Florida (UCF) in the U.S. He owns those distinctions which are the selectively awarded titles, such as the elected member of the European Academy of Sciences (M.EAS), the Board Certified Environmental Engineer (BCEE, formerly DEE), Diplomat of Water Resources Engineer (D.WRE), and Certificate of Leadership in Energy and Environment Design (LEED). He is also members of 12 professional associations. He was one of the founding fellows of the International Society of Environmental Information Management in 2002. In recent years, the focus of his research brings well-rounded interdisciplinary efforts in the area of environmental informatics and systems analysis. It emphasizes fusion of environmental hydrology, environmental/ecological engineering processes, computational methods, and information technologies to advance our understanding of large, complex, and integrated environmental and hydrologic systems. He has authored and co-authored over 130 peer-reviewed journal articles, 9 books and 7 book chapters, and 128 conference papers with more than 1,000 citations. He served as the guest editor for seven special issues and also serves on the editorial board of 11 international journals and ad hoc reviewers of 68 international journals.

Dr. Robert Clark, P.E., NRMRL-UC WRAP Team
Robert Clark received a B.S. Degree in Civil Engineering from Oregon State University (1960), a B.S. Degree in Mathematics from Portland State University (1961), an M.S. in Mathematics from Xavier University (1964), an M.S. in Civil Engineering from Cornell University (1968) and a Ph.D. in Environmental Engineering from the University of Cincinnati (1976). Dr. Clark is a registered engineer in the State of Ohio and worked as an environmental engineer in the U.S. Public Health Service and the EPA from 1961 to August 2002. He was Director of the EPA's Water Supply and Water Resources Division (WSWRD) in the Office of Research and Development's (ORD) National Risk Management Research Laboratory (NRMRL) for fourteen years (1985-1999). In 1999 he was appointed to a Senior Expert Position in EPA with the title Senior Research Engineering Advisor (1999-2002). He retired from EPA in August of 2002 and is now an independent consultant. He is an Adjunct Professor of Civil and Environmental Engineering at the University of Cincinnati and recently completed service as a member of the National Research Council's Committee on "Public Water Distribution Systems: Assessing and Reducing Risks." He has published over 375 papers and five books and is a life member of the American Water Works Association (AWWA) and the American

Proceedings of the First National Expert and Stakeholder Workshop on
Water Infrastructure Sustainability and Adaptation to Climate Change

115

Society of Civil Engineers (ASCE). He has served on numerous professional and technical committees and is currently a member the Water Supply Working Group (WSWG) for the Greater Cincinnati Water Works (GCWW). The WSWR was convened by the City Manager for the City of Cincinnati to evaluate the possibility of converting the GCWW to a regional water district. He has received numerous awards for his work including: the Environmental and Water Resources Institute's (American Society of Civil Engineers) Best Research Paper Award from the Journal of Water resources Planning and Management for 2006, the American Society of Civil Engineers (Environmental and Water Resources Institute) Lifetime Achievement Award for 2004 in recognition of a life-long and eminent contribution to the environmental and water resources engineering disciplines through practice, research and public service, and the U.S. EPA Distinguished Service Career Award for lifetime accomplishments, and leadership as a researcher and manager in the field of water supply (2002).

Carol Collier, Executive Director, Delaware River Basin Commission
Carol Collier was appointed Executive Director of the Delaware River Basin Commission (DRBC) on August 31, 1998. The DRBC is an interstate/federal commission that provides a unified approach to water resource management without regard to political boundaries. Before joining DRBC, Ms. Collier was Executive Director of Pennsylvania's 21st Century Environment Commission. Governor Tom Ridge formed the Environment Commission in 1997 to establish the Commonwealth's environmental priorities and to recommend a course of action for the next century. At the time Governor Ridge asked Ms. Collier to serve as executive director for the 21st Century Environment Commission, she was Regional Director of the Pennsylvania Department of Environmental Protection (PADEP) Southeast Region. Prior to PADEP, she served 19 years with BCM Environmental Engineers, Inc., Plymouth Meeting, Pa., beginning as a student intern and ultimately becoming Vice President of Environmental Planning, Science and Risk. She has a B.A. in Biology from Smith College and a Masters in Regional Planning from the University of Pennsylvania. She is a Professional Planner licensed in the State of New Jersey, a member of the American Institute of Certified Planners (AICP) and a Certified Senior Ecologist. She is a member of her township's environmental protection advisory board, on the Boards of the American Water Resources Association (AWRA) and the newly formed Clean Water America Alliance (CWAA), teaches environmental management courses at the University of Pennsylvania and has published on environmental and water-related topics.

Elizabeth Corr, EPA Office of Water
Elizabeth Corr has been the Associate Director of the Drinking Water Protection Division in the Office of Ground Water and Drinking Water, U.S. Environmental Protection Agency, since 2001. Prior, Ms. Corr served as Special Assistant for water issues in EPA's Office of the Administrator; led a team to develop drinking water treatment regulations, for which she received EPA's Gold Medal; and worked with states to protect ground water. She began her career in Washington DC as staff to the Subcommittee on Transportation and Hazardous Materials in the U.S. House of Representatives in 1987.

John Cromwell, Stratus Consulting, Inc.

John Cromwell is an environmental economist with Stratus Consulting. Mr. Cromwell has over 30 years of experience specializing in the water and wastewater utility sector. In addition to recent work in climate change, he has analyzed a broad range of national policy issues affecting the water sector, including: costs and benefits of regulations governing water quality, infrastructure rehabilitation and replacement investments, regional collaboration schemes, stormwater management initiatives, combined sewer overflows, and utility management and financial planning issues. Based in Washington, DC, he has been centrally involved in national policy issues affecting the water industry as an advisor to Congress, federal agencies, state regulators, and industry research organizations. He holds BS degrees in biology and economics as well as a Master of Policy Sciences degree from the University of Maryland.

Dennis Diemer, East Bay MUD, CA

Dennis Diemer, General Manager of East Bay Municipal Utility District (EBMUD), has over 30 years of experience with public agencies and engineering consulting firms in the planning, design and operation of water and wastewater systems. Mr. Diemer has served as EBMUD's General Manager since 1995. He holds a BS in Civil Engineering from Loyola University in Los Angeles, and a Masters in Civil and Environmental Engineering from Stanford University. He is a registered Civil Engineer with the State of California. He is an active member of the Association of California Water Agencies (ACWA), American Water Works Association (AWWA), Water Environment Federation, and California Urban Water Agencies. He currently serves on EPA's National Drinking Water Advisory Council, AWWA's Water Utility Council, the Water Education Foundation, and is the current Chairman of the Water Environment Research Foundation.

Cynthia Dougherty, Director, EPA Office of Ground Water and Drinking Water

Cynthia Dougherty is the Director of the Office of Ground Water and Drinking Water at the U.S. Environmental Protection Agency in Washington, DC. In that capacity, Ms. Dougherty serves as EPA's national program manager for implementation of the federal Safe Drinking Water Act. Prior to her current position, she served as the Director of the Permits Division in the Office of Wastewater Management. She has also served in EPA's Office of Enforcement and Office of Planning and Management. She has a degree from Duke University and is the recipient of three Presidential Meritorious Executive Awards for her federal service.

Dr. David Easterling, NOAA National Climatic Data Center

David Easterling is currently Chief of the Scientific Services Division at NOAA's National Climatic Data Center in Asheville, NC. Dr. Easterling received his Ph.D. from the University of North Carolina at Chapel Hill in 1987 and served as an Assistant Professor in the Atmospheric Sciences Program, Department of Geography, at the University of Indiana-Bloomington from 1987 to 1990. In 1990, he moved to the National Climatic Data Center as a research scientist, was appointed Principal Scientist in 1999, and Chief of Scientific Services in 2002. He has authored or co-authored more than eighty research articles on climate change issues in journals such as Science, Nature and the Journal of Climate. He was a Lead Author for the Nobel Prize winning Intergovernmental Panel on Climate Change (IPCC) Fourth Assessment Report, a Convening Lead Author for the U.S. Climate Change Science Program (CCSP) Synthesis and Assessment Product (SAP) 3.3 on Climate Extremes and was a Contributing Author to the IPCC Second and Third Assessment Reports. He is a Fellow of the American Meteorological Society and his research interests include the detection of climate change in the observed record, particularly changes in extreme climate events and the assessment of climate model simulations for changes in extreme climate events.

Proceedings of the First National Expert and Stakeholder Workshop on
Water Infrastructure Sustainability and Adaptation to Climate Change

117

Mikaela Engert, City of Keene, New Hampshire

Mikaela Engert works as a city planner for the City of Keene, New Hampshire. Much of Ms. Engert's professional experience and education in the field of planning focuses on community sustainability issues. Specifically, her interests are in food security, climate change, and open space planning. She earned a Master's of Urban Planning from the State University of New York at Buffalo and obtained her Bachelor's from Green Mountain College in Vermont. She was also part of a team of graduate students which earned the "Outstanding Student Project" awards from both the APA Western New York Section, the APA New York Upstate Chapter, as well as from the American Institute of Certified Planners for the plan entitled: Food for Growth: A Community Food System Plan for Buffalo's West Side. She currently guides the City of Keene's climate change and long-term sustainability initiatives. In the summer of 2006, the City of Keene was selected as a pilot community to test and evaluate ICLEI's latest program, Climate Resilient Communities (CRC). The CRC program seeks to assist municipalities in planning for the predicted impacts associated with global climate change in order to improve a community's long-term preparedness for climate impacts. She led the team through the CRC process to create one of the first municipal adaptation plans in the country.

Stephen Estes-Smargiassi, Massachusetts Water Resources Authority

Stephen Estes-Smargiassi is a planner and an engineer with an interest in complex multi-disciplinary projects. In his over 20 years at the MWRA, Mr. Estes-Smargiassi has led or participated in all drinking water quality and master planning initiatives. He is active with the AWWA Research Foundation (now Water Research Foundation), is a QualServe peer review team leader, and has actively participated in water quality regulatory development activities regionally and nationally. As part of his responsibilities he oversaw and evaluated the MWRA's successful demand management programs, reducing water demand by about one-third; initiated its GIS system; and coordinated protection planning studies for MWRA's watersheds, as well as for about 40 other smaller supply systems in the Boston metropolitan area. His group recently completed an integrated water and wastewater master plan to prioritize and schedule improvements to the region's water and sewer systems over the next 20 years. Over the past few years, his priorities have been developing the briefing materials used by MWRA's Board of Directors to make the treatment technology decision for the metropolitan Boston water system and then participating in the successful defense of that decision in federal court; producing and distributing the MWRA's annual water quality report to over 800,000 households; and using the opportunity of both processes to reinforce the bridges built over the past decade to the public health community. He is currently overseeing drinking water quality and public health outcome research to understand and evaluate recent treatment improvements. He has been involved in thinking about how water and wastewater systems can adapt to climate change since EPA, the Army Corps and other agencies called together the First National Conference on Climate Change and Water Resources Management in 1991. He continues to conduct research and policy development activities at MWRA and regionally on how the changing climate ought to affect planning and investment decisions. He has a Bachelor's of Civil Engineering from the Massachusetts Institute of Technology and a Master's in City and Regional Planning from Harvard University. He lives in Boston where the streets do not follow old cowpaths, although they seem to, loves maps, and has two kids who also love maps. And he proudly drinks tap water, especially in Boston.

Josh Foster, Center for Clean Air Policy

Josh Foster manages CCAP's Urban Leaders Adaptation Initiative, designed to equip U.S. partner cities and counties make effective policy and investment decisions to increase their resiliency to the impacts of climate change. Mr. Foster has 13 years of experience working on climate adaptation at the National Oceanic and Atmospheric Administration (NOAA) Climate Science Program Office as a manager for climate research applications and services. His work focused on decision support, drought and water resources management, local urban preparedness, and engagement with the

Proceedings of the First National Expert and Stakeholder Workshop on
Water Infrastructure Sustainability and Adaptation to Climate Change

118

private sector. He was the project manager for NOAA's Climate Resilient Communities project from 2005-08 in collaboration with ICLEI-Local Governments for Sustainability. In the past he has also worked on NOAA's Regional Integrated Climate Sciences and Assessments (RISA) Program, the International Research Institute for Climate and Society (IRI), the Nobel Prize winning Intergovernmental Panel on Climate Change (IPCC), the United Nations Development Program, and the White House Office on Environmental Policy. He holds a Master's in International Relations and Environmental Management from Yale University, and a Bachelor's in International Relations and Environmental Policy with a Minor in Latin American Studies from the University of Massachusetts at Amherst.

Dr. Alice Gilliland, EPA National Exposure Research Laboratory

Alice Gilliland is from the EPA Office of Research and Development (ORD), and she is located in Research Triangle Park, NC. In the Atmospheric Modeling Division, Dr. Gilliland is chief of the Applied Modeling Research Branch. She earned her Ph.D. from the Georgia Institute of Technology in the field of atmospheric sciences. Her areas of expertise include climate influences on air quality, regional scale air quality modeling, and evaluation of airborne emissions and models. Over the past five years, she has led the ORD Climate Impacts on Regional Air Quality (CIRAQ) project. CIRAQ has contributed to a USEPA Global Change Research Program interim report (EPA/600/R-07/094) on ozone impacts from future climate, the Climate Change Science Program (CCSP) Synthesis and Assessment Product 3.2, and several recent journal articles. As the importance of climate impacts on air quality and ecosystems increases in priority for EPA, she is expanding the CIRAQ program with regional climate downscaling capabilities to extend the existing meteorological modeling expertise in the Division. The regional climate scenarios can be used for assessments of air quality, water quantity and quality, and other ecosystem issues.

Dr. Peter Gleick, Pacific Institute

Peter Gleick is co-founder and President of the Pacific Institute in Oakland, California. The Institute is one of the world's leading non-partisan policy research groups addressing global environment and development problems, especially in the area of freshwater resources. Dr. Gleick is an internationally recognized water expert. His research and writing address the hydrologic impacts of climate change, sustainable water use, water privatization, and international conflicts over water resources. His work on sustainable management and use of water led to him being named by the BBC as a "visionary on the environment" in its Essential Guide to the 21st Century. In 2008, Wired Magazine called him "one of 15 People the Next President Should Listen To." He is one of the nation's leading scientists working on the implications of climate change for water resources. He has also played a leading role in highlighting the risks to national and international security from conflicts over shared water resources. He produced some of the earliest assessments of the connections between water and political disputes and has briefed major international policy makers ranging from the Vice President and Secretary of State of the United States to the Prime Minister of Jordan on these issues. He also has testified regularly for the U.S. Senate, House of Representatives, and state legislatures, and briefed international governments and policy makers. He received a B.S. from Yale University and an M.S. and Ph.D. from the University of California, Berkeley. In 2003 he received a MacArthur Foundation Fellowship for his work on global freshwater issues. He was elected an Academician of the International Water Academy, in Oslo, Norway, in 1999. In 2006 he was elected to the U.S. National Academy of Sciences, Washington, D.C. and his public service includes work with a wide range of science advisory boards, editorial boards, and other organizations. He is the author of more than 80 peer-reviewed papers and book chapters, and six books, including the biennial water report The World's Water published by Island Press (Washington, D.C.).

Proceedings of the First National Expert and Stakeholder Workshop on
Water Infrastructure Sustainability and Adaptation to Climate Change

119

Dr. James Goodrich, EPA ORD

James Goodrich has been employed by the EPA, Office of Research and Development (ORD) for 32 years. Dr. Goodrich has a Ph.D. and B.S. from the University of Cincinnati and an M.S. from Florida State University. He has managed large multidisciplinary research programs relative to drinking water, wastewater, and watershed management and has authored or co-authored several peer reviewed journal articles, EPA Handbooks, and book chapters. Currently, he is involved in the development and implementation of the Water Resource Adaptation to Climate Change Program in ORD.

Dr. Walter Grayman, P.E., NRMRL-UC WRAP Team

For the past 25 years, Walter Grayman has been owner of the independent consulting engineering firm of W.M. Grayman Consulting Engineer in Cincinnati, Ohio. Dr. Grayman holds a PhD and SM degree in civil engineering (specializing in water resources) from MIT and a BS degree in civil engineering from Carnegie Mellon University. He is a registered Professional Engineer in the State of Ohio, a Diplomat of the American Academy of Water Resources Engineers, and an Adjunct Professor in the Civil and Environmental Engineering Department at the University of Cincinnati. He is a member of ASCE, AWWA, AGU, IWA and EWRI, and has chaired national committees in both AWWA and ASCE. He has over 130 publications including co-author of the AWWA book Modeling Water Quality in Drinking Water Distribution Systems and contributing author for McGraw Hill Handbooks on Water Distribution Systems and Water Supply Systems Security. He has won awards from both ASCE and AWWA for his publications, research and service to the profession. He has specialized in the development and application of models and quantitative analysis within the field of water resources. He has performed pioneering work in the areas of managing, sampling, analyzing and modeling hydraulics and water quality in water distribution systems and storage tanks. He has been actively involved in the area of water system security as a certified RAM-W trainer, in the preparation of vulnerability assessments and in research on contamination of water systems for EPA, CDC, AwwaRF and other organizations. He has worked extensively in the field of spatial data analysis and geographic information system technology for over three decades including the integration of hydrologic, riverine and infrastructure models with GIS technology. As a consultant to the United Nations Industrial Development Organization (UNIDO), he has served as an international consultant on pollution prevention and stream modeling projects in Turkey, Vietnam and Ecuador.

Benjamin Grumbles, Assistant Administrator, EPA Office of Water

Benjamin H. Grumbles was confirmed by the United States Senate on November 20, 2004, as Assistant Administrator for Water at the U.S. Environmental Protection Agency. Prior to that Mr. Grumbles served as Deputy Assistant Administrator for Water and Acting Associate Administrator for Congressional and Intergovernmental Relations. Before coming to EPA in 2002, he was Deputy Chief of Staff and Environmental Counsel for the Committee on Science in the U.S. House of Representatives. He also served for over 15 years in various capacities on the House Transportation and Infrastructure Committee, including Senior Counsel for the Water Resources and Environment Subcommittee. From 1993 to 2004, he was an adjunct professor of law at the George Washington University Law School, teaching courses on the Clean Water Act, Safe Drinking Water Act, Ocean Dumping Act, and Oil Pollution Act. His degrees include a B.A., Wake Forest University; J.D., Emory University; and LL.M. in Environmental Law, from the George Washington University Law School. He was born and raised in Louisville, Kentucky. He lives in Arlington, Virginia, with his wife, Karen and their two water-loving kids.

Sally Gutierrez, Director, EPA ORD National Risk Management Research Laboratory

Sally C. Gutierrez is the Director of the National Risk Management Research Laboratory (NRMRL) in Cincinnati, Ohio. NRMRL is one of three Federal research laboratories within the U.S. Environmental Protection Agency's Office of Research and Development. The Laboratory is responsible for

Proceedings of the First National Expert and Stakeholder Workshop on
Water Infrastructure Sustainability and Adaptation to Climate Change

120

conducting engineering and environmental technology research to support the Agency in development of policy, regulations and guidance to further environmental protection in the U.S. The research staff consists of 400 environmental and chemical engineers, chemists, microbiologists, economists, hydrologists and other scientists and support staff. Key areas of research include: treatment and control of contaminants in drinking water, restoration of ecosystems, control of air pollutants, remediation of contaminated sites, environmental sustainability and environmental technology testing and development.

Ms. Gutierrez was born and raised in Houston, Texas. She received a Master of Science degree from the University of Texas, School of Public Health in Houston. Her area of expertise is water resource management. She was appointed NRMRL's Director in 2005. Prior to this appointment she was the Director of the Water Supply and Water Resources Division with the Laboratory. During her tenure as Director of the Water Supply and Water Resources Division, she was responsible for leading a national technology demonstration program for control of arsenic in drinking water. Prior to coming to EPA, she was responsible for administering several water programs for the State of Texas environmental agency in the areas of drinking water, water monitoring, wastewater treatment permitting, and utility rates. She is a member of the American Water Works Association and the American Society of Civil Engineers and is past President of the Texas Environmental Health Association. She is a member of the Board of Directors for AIDIS U.S.A.

James A. Hanlon, P.E., Director, EPA Office of Wastewater Management
James A. Hanlon was appointed Director of the Office of Wastewater Management (OWM) in the Office of Water in April 2002. OWM is responsible for the management of the NPDES program which permits municipal and industrial wastewater discharges, and the administration of Federal financial and technical assistance for publicly owned wastewater treatment works. Mr. Hanlon is a career civil servant with over 30 years of government service with the Environmental Protection Agency (EPA). In 1984, he was appointed to the position of Director, Municipal Construction Division, and was responsible for the management of EPA's national construction grants and state revolving fund programs, providing assistance to municipalities in their wastewater infrastructure construction programs. He was appointed to the position of Deputy Director of the Office of Science and Technology in the Office of Water in 1991. In this capacity, he was responsible for the scientific and technical basis of the federal water quality and safe drinking water programs. From January 2001 to April 2002, he served as Acting Deputy Assistant Administrator for the Office of Water. He earned a Bachelor of Science Degree in Civil Engineering from the University of Illinois and a Master of Business Administration Degree from the University of Chicago. He is also a registered Professional Engineer in the State of Illinois.

Roy Haught, EPA National Risk Management Research Laboratory
Roy C. Haught has over 20 years of diversified and professional management experience in the design, fabrication, testing, and evaluation of various research and development (R&D) innovative treatment technologies. He is currently the acting chief of the Water Quality Management Branch, National Risk Management Research Laboratory, Officer of Research and Development.

Dr. Robert Hirsch, U.S. Geological Survey
Robert M. Hirsch currently serves as a Research Hydrologist at the USGS. From 1994 through May 2008, he served as the Chief Hydrologist of the U.S. Geological Survey (in the later years of his tenure the position was titled Associate Director for Water). In this capacity, Dr. Hirsch was responsible for all U.S. Geological Survey (USGS) water science programs. These programs encompass research and monitoring of the nation's ground water and surface water resources including issues of water quantity as well as quality. Since 2003 he has served as the co-chair of the Subcommittee on Water Availability and Quality of the Committee on Environment and Natural

Resources of the National Science and Technology Council. He began his USGS career in 1976 as a hydrologist and has conducted research on water supply, water quality, pollutant transport, and flood frequency analysis. He had a leading role in the development of several major USGS programs: 1) the National Water Quality Assessment (NAWQA) Program: 2) the National Streamflow Information Program (NSIP); and 3) the National Water Information System Web (NWISWeb). Previous leadership positions include: Acting Director of the USGS during an interim period between Directors (August 1993 to March 1994); Assistant Chief Hydrologist for Research and External Coordination (1989-1993); and Staff Assistant to the Assistant Secretary for Water and Science, U.S. Department of the Interior (1987-1988). He received degrees from Earlham College (bachelor of arts degree in geology, 1971), University of Washington (master of sciences degree in geology, 1974), and The Johns Hopkins University (doctorate in geography and environmental engineering, 1976). He has received numerous honors from the Federal Government and from non-governmental organizations, including the 2006 American Water Resources Association's William C. Ackermann Medal for Excellence in Water Management, and has twice been conferred the rank of Meritorious Senior Executive by the President of the United States. He is co-author of the textbook "Statistical Methods in Water Resources." He is a Fellow of the American Association for the Advancement of Science and an active member of the American Geophysical Union and the American Water Resources Association.

Dr. Thomas Johnson, EPA National Center for Environmental Assessment

Thomas Johnson is a hydrologist with the EPA, Office of Research and Development, Global Change Research Program. Dr. Johnson's research interests include the assessment and management of climate and land use change impacts on water and watershed systems, documenting and improving the effectiveness of stream and watershed restoration, and the development of decision support tools for adapting to climate change. Prior to joining EPA he held positions with the Academy of Natural Sciences of Philadelphia and was an AAAS Science and Technology Policy Fellow in Washington, DC. He has degrees from the University of Colorado (B.A. Environmental Biology), Colorado State University (M.S. Watershed Sciences), and Penn State University (Ph.D. Forest Hydrology).

Dr. Paul Kirshen, Tufts University

Since 1996, Paul Kirshen has been a Research Professor in the Civil and Environmental Engineering Department of Tufts University. Dr. Kirshen is also Affiliated Faculty in the Department of Urban and Environmental Policy and Planning and Adjunct Professor in the Friedman School of Nutrition Science and Policy. Since 2004, he has been Director and co-founder of the Tufts Water: Systems, Science, and Society (WSSS) Interdisciplinary Graduate Education Program. He is also presently a UCOWR Fellow in support of the U.S. Army Corps of Engineers Institute for Water Resources in Integrated Water Resources Management. He also recently joined Battelle Memorial Institute in a part time capacity as a Research Leader, and will assume a full time role with Battelle in June 2009. He is an expert in climate change impacts and adaptation, and integrated water resources management. He has carried out analyses of climate change impacts on water resources systems and adaptation actions on global, national, and local scales. He was Principal Investigator of one of the first integrated, risk-based assessments of climate change impacts and adaptation options for urban infrastructure systems, the metro Boston CLIMB study. He was also the lead of the coastal flooding team for the Union of Concerned Scientists' Northeast Climate Impacts Assessment Report. Presently he is leading a NOAA funded effort to develop an integrated, scenario-based, risk assessment procedure for urban drainage management under a changing climate. He is also involved in research on climate change and environmental justice. His primary focus for international research is West Africa where he has been working with a team for many years on the use of seasonal climate forecasting to improve agriculture and water management, a climate change adaptation strategy (CFAR project). He received both his doctorate and master's degree in holds

Proceedings of the First National Expert and Stakeholder Workshop on
Water Infrastructure Sustainability and Adaptation to Climate Change

122

Civil Engineering from Massachusetts Institute of Technology and a bachelor's degree in engineering from Brown University.

Dr. Dennis Lettenmaier, University of Washington

Dennis Lettenmaier received his B.S. in Mechanical Engineering (summa cum laude) at the University of Washington in 1971, his M.S. in Civil, Mechanical, and Environmental Engineering at the George Washington University in 1973, and his Ph.D. at the University of Washington in 1975. He joined the University of Washington faculty in 1976. In addition to his service at the University of Washington, he spent a year as visiting scientist at the U.S. Geological Survey in Reston, VA (1985-86) and was the Program Manager of NASA's Land Surface Hydrology Program at NASA Headquarters in 1997-98. He is a member of the American Geophysical Union, the American Water Resources Association, the European Geosciences Union, the American Meteorological Society, and the American Society of Civil Engineers, and the American Association for the Advancement of Science. He was a recipient of ASCE's Huber Research Prize in 1990, and the American Geophysical Union's Hydrology Section Award in 2000. He is a Fellow of the American Geophysical Union, the American Meteorological Society, and the American Association for the Advancement of Science, and is a member of the International Water Academy. He is an author or co-author of over 200 journal articles. He was the first Chief Editor of the American Meteorological Society Journal of Hydrometeorology, and is currently an Associate Editor of Water Resources Research. He is the President-elect of the Hydrology Section of the American Geophysical Union. His areas of research interest are large scale hydrology, hydrologic aspects of remote sensing, and hydrology-climate interactions.

Dr. Audrey Levine, P.E., EPA Office of Research and Development

Audrey Levine is the National Program Director for Drinking Water at EPA. Dr. Levine is an environmental engineer with extensive research experience in water quality, water treatment and distribution systems, treatment technologies, and water reuse. Prior to joining EPA, she was a faculty member of the Department of Civil and Environmental Engineering at the University of South Florida in Tampa. She is a Diplomat of Environmental Engineering (DEE) and a registered professional engineer (P.E.). She has more than 20 years of broad-based, technical experience within academic, government, industry, and consulting settings. She has a doctorate in civil engineering from the University of California at Davis, and a master's degree in Public Health from Tulane University.

Dr. David Major, Columbia University

David Major is Senior Research Scientist at the Columbia University Earth Institute's Center for Climate Systems Research. Dr. Major completed his undergraduate work at Wesleyan University and the London School of Economics, and received the Ph.D. in Economics from Harvard. He has been a faculty member at MIT and at Clark University, a Visiting Fellow at Clare Hall, Cambridge, a senior planner with the New York City Water Supply System, and Program Director for Global Environmental Change at the Social Science Research Council. His principal scientific research focus at Columbia is the adaptation of urban infrastructure to global climate change. He is the award-winning author, co-author or co-editor of twelve books on natural resources planning, environmental management, biography and literary studies.

Linda Mearns, National Center for Atmospheric Research

Linda Mearns is the Director of the Weather and Climate Impacts Assessment Science Program (WCIASP) within the Institute for the Study of Society and the Environment (ISSE) and Senior Scientist at the National Center for Atmospheric Research, Boulder, Colorado. Dr. Mearns served as Director of ISSE for three years ending in April 2008. She holds a Ph.D. in Geography/Climatology from UCLA. She has performed research and published mainly in the areas of climate change

Proceedings of the First National Expert and Stakeholder Workshop on
Water Infrastructure Sustainability and Adaptation to Climate Change

123

scenario formation, quantifying uncertainties, and climate change impacts on agro-ecosystems. She has particularly worked extensively with regional climate models. She has most recently published papers on the effect of uncertainty in climate change scenarios on agricultural and economic impacts of climate change, and quantifying uncertainty of regional climate change. She has been an author in the IPCC Climate Change 1995, 2001, and 2007 Assessments regarding climate variability, impacts of climate change on agriculture, regional projections of climate change, climate scenarios, and uncertainty in future projections of climate change. For the 2007 Report(s) she was Lead Author for the chapter on Regional Projections of Climate Change in Working Group 1 and for the chapter on New Assessment Methods in Working Group 2. She is also an author on two Synthesis Products of the US Climate Change Science Program. She leads the multi-agency supported North American Regional Climate Change Assessment Program (NARCCAP), which is providing multiple high-resolution climate change scenarios for the North American impacts community. She is a member of the National Research Council Climate Research Committee (CRC) and Human Dimensions of Global Change (HDGC) Committee. She was made a Fellow of the American Meteorological Society in January 2006.

Karen Metchis, EPA OW Office of Wastewater Management
Karen Metchis works for the EPA Office of Water in Washington, D.C. Ms. Metchis began her career at EPA in 1992, and has worked on various projects, including implementing the Montreal Protocol on Substances that Deplete the Ozone Layer, protecting the Florida Everglades from urban encroachment, and developing regulations such as the Concentrated Animal Feeding Operations rule. She has been in the Office of Wastewater Management for the past 10 years, and is an active member of the EPA Office of Water's Climate Workgroup. In addition to co-planning this workshop, she is currently working as the Office of Water Transition Coordinator.

Dan Murray, P.E., EPA National Risk Management Research Laboratory
Dan Murray is a Senior Environmental Engineer with the EPA Office of Research and Development (ORD) in Cincinnati, Ohio, and has been with EPA for over 28 years. Mr. Murray is currently leading EPA's Aging Water Infrastructure Research Program. He received his BS in Civil Engineering from Merrimack College in North Andover, Massachusetts and his MS in Civil Engineering from Northeastern University in Boston, Massachusetts. Prior to joining ORD, he worked in EPA Region 1 in Boston, and EPA Region 5 in Cleveland. He also worked for the Massachusetts Water Resources Authority, leading the CSO control program. In 1995, he received the Gold Medal for Exceptional Service, EPA's highest honor, for his work in supporting the development of the Agency's CSO Policy. He is a registered Professional Engineer in Massachusetts and Ohio and an active member of the Water Environment Federation and the American Society of Civil Engineers.

Chuck Noss, EPA Office of Research and Development
Chuck Noss is the National Program Director for Water Quality at EPA and brings a strong and diverse scientific background to the water quality research program. He has extensive water quality expertise in wastewater collection and treatment systems, stormwater management, and environmental impacts. Prior to joining EPA, he served as deputy executive director and director for research at the Water Environmental Research Foundation. He has a doctorate in science in environmental health engineering from The Johns Hopkins University.

Proceedings of the First National Expert and Stakeholder Workshop on
Water Infrastructure Sustainability and Adaptation to Climate Change

124

Doug Owen, P.E., Malcolm Pirnie Inc.

Douglas M. Owen, P.E., BCEE is a Vice President with Malcolm Pirnie, Inc. and is the Chief Technology Officer for the firm. In that role, Mr. Owen is responsible for technology applications for clients and the firm, applied research programs, outreach to universities, and knowledge management. Prior to 2007, he served as Managing Director of Malcolm Pirnie's business unit providing drinking water, wastewater, and water resources services to municipal clients. He has specialized in water and wastewater planning and design since he received his Bachelor's Degree in Civil Engineering from Purdue University in 1980 and his Master's Degree in Environmental Sciences and Engineering from the University of North Carolina at Chapel Hill in 1982. He has led applied research projects on advanced technologies, consulted with utilities on treatment and facility planning for over 6 billion gallons per day of treatment capacity throughout the United States, and has provided technical and facilitation support to USEPA and AWWA on a range of policy issues since 1990 - including drinking water regulatory development, advanced technology implementation and utility compliance. He is currently the Chair of the Editorial Advisory Board for the Journal of the American Water Works Association, serves on EPA's National Drinking Water Advisory Council, has served as a Trustee for AWWA's Water Science and Research Division, and serves on advisory boards for the University of Texas, Columbia University, and the University of North Carolina School of Public Health. He has published widely on water planning and design topics in books, peer-reviewed journals, and at national and international conferences.

David Rager, Greater Cincinnati Water Works

As Director of the Greater Cincinnati Water Works, David Rager oversees a utility that serves approximately 1,000,000 people over about 800 square miles in southwestern Ohio and northern Kentucky. He has worked to develop a strategic business plan, utilize employee work teams, have regular customer surveys and focus groups for insight into service delivery, expand into new services and service areas, and use technology to manage costs and activity-based resource management, and work in innovative ways with other utilities in the region to solve regional water supply and utility services issues.

Nationally, he serves on the Board of Directors for the Association of Metropolitan Water Agencies where he recently served as President, the American Water Works Association Manufacturers/Associates Council, and the Water Utility Council. In 2007, the Awwa Research Foundation (AwwaRF), the leading nonprofit water research foundation dedicated to advancing the science of drinking water, elected Mr. Rager as chairman of the board for the term of 2007 - 2010.

Paul Rush, P.E., NYC Bureau of Water Supply

Paul Rush presently serves as Deputy Commissioner for the New York City Department of Environmental Protection's (DEP) Bureau of Water Supply and is responsible for operating and protecting New York City's upstate water supply system in order to deliver sufficient high quality drinking water to the 9 million New York State residents in nine counties who rely on the City's system. Until October 2006, Mr. Rush served as Director of the West of Hudson (WOH) Operations Division and was responsible for the operation and maintenance of all New York City water supply & wastewater treatment facilities west of the Hudson River. In prior assignments, he served as the Delaware District Engineer and the Delaware District Operations Chief where he focused solely on the operation of the City's Delaware Water Supply System. He has worked for the New York City Department of Environmental Protection since 1992. Prior to his employment with New York City he served on active duty in the Army as an Engineer Officer. He holds a Master of Science degree in Civil Engineering from Michigan Technological University and Bachelor of Science degree in Civil Engineering from the United States Military Academy. He is a registered professional engineer in the state of New York.

Proceedings of the First National Expert and Stakeholder Workshop on
Water Infrastructure Sustainability and Adaptation to Climate Change

125

Dr. Joel Scheraga, EPA Office of Research and Development
Joel Scheraga is the National Program Director for the Global Change Research Program and the Mercury Research Program in the U.S. Environmental Protection Agency's Office of Research and Development. Dr. Scheraga is responsible for managing a $20.0 million Global Change Research Program, a $4 million Mercury Research Program, and over 40 personnel in five laboratories and centers. He is also the EPA Principal Representative to the U.S. Climate Change Science Program (CCSP), which coordinates and integrates scientific research on climate and global change supported by the U.S. Government. He has participated in the Intergovernmental Panel on Climate Change (IPCC), which was awarded the 2007 Nobel Peace Prize. He was Chair of the U.S. Global Change Research Program's National Assessment Workgroup from 2000-2002 and Vice Chair from 1998-2000. The Workgroup was responsible for managing the U.S. National Assessment process which resulted in the report to Congress entitled, "Climate Change Impacts on the United States: The Potential Consequences of Climate Variability and Change." He was a co-author of the 2005 Human Health Synthesis Report that is part of the Millennium Ecosystem Assessment. He has served as a faculty member for the International Water Management Course held by the Swiss Federal Institute of Environmental Science and Technology in Switzerland. He was a co-editor and lead author of the book, Climate Change and Human Health: Risks and Responses, released by the World Health Organization in December 2003, and co-author of the 2003 WHO report, *Methods of Assessing Human Vulnerability and Public Health Adaptation to Climate Change.* He co-authored a white paper in 2003 on the effects of climate change on water quality in the Great Lakes Region for the US/Canada International Joint Commission's Water Quality Board.

Dr. Scheraga received an A.B. degree in geology-mathematics/physics from Brown University in 1976, an M.A. in economics from Brown University in 1979, and a Ph.D. in economics from Brown University in 1981. Prior to joining EPA, he was an Assistant Professor of Economics at Rutgers University from 1981-1987, and a Visiting Assistant Professor of Economics at Princeton University from 1985-1986. He was named a Fellow of the Institute for Science, Technology and Public Policy in The Bush School of Government and Public Service at Texas A&M University in June 2008. He was also the recipient of the 2004 inaugural Horace Mann Distinguished Graduate School Alumni Award presented by Brown University. He was one of the 1,360 scientists from 95 countries honored with the 2005 Zayed Award for scientific and/or technological achievement in environment for their work on the Millennium Ecosystem Assessment. He has also received six EPA Bronze Medals.

Dr. Michael Shapiro, Deputy Assistant Administrator, EPA Office of Water
Michael Shapiro joined the Office of Water as the Deputy Assistant Administrator in November 2002. Prior to that, Dr. Shapiro was the Principal Deputy Assistant Administrator for the Office of Solid Waste and Emergency Response (OSWER). He has been in that position since February 1997, with a brief nine months as Acting Assistant Administrator during the transition between Administrations. Before that he was the Director of the Office of Solid Waste, where he had served since November 1993. Prior to that, he served first as Deputy Assistant Administrator and then as Acting Assistant Administrator in EPA's Office of Air and Radiation, where he directed implementation of the 1990 Clean Air Act Amendments. From 1980 to 1989, he held a variety of positions in the Office of Pesticides and Toxic Substances, where one of his responsibilities was developing EPA's Toxic Release Inventory. He has a B.S. in Mechanical Engineering from Lehigh and a Ph.D. in Environmental Engineering from Harvard. He has also taught in the public policy program at the John F. Kennedy School of Government.

Dr. Daniel Sheer, HydroLogics, Inc.
Daniel Sheer is the founder and President of HydroLogics. Dr. Sheer has devoted his professional career to improving water management. After receiving his Ph.D with honors from the Johns Hopkins University in 1974, he became the Planning Engineer, and then the Technical Director of

Proceedings of the First National Expert and Stakeholder Workshop on
Water Infrastructure Sustainability and Adaptation to Climate Change

126

the Interstate Commission on the Potomac River Basin. In these capacities, he designed the technical work plan for the Washington Metropolitan Area 208 Plan (water quality management), and led the technical development effort that provided a long term water supply solution for the same region. He was the first Director of CO-OP, the new institution designated to implement that plan.

In 1985 Dr. Sheer founded Water Resources Management, Inc., now renamed HydroLogics. He has been directly involved in the majority of HydroLogics' projects, and was instrumental in the creation of the Southern Nevada Water Authority and the Kansas River Water Assurance District. For the past decade, Dr. Sheer has been closely involved in planning and operations for the Everglades, Lake Okeechobee, the Everglades Agricultural Area and the Lower East Coast through contracts with the South Florida Water Management District. He has directed the modeling of the Delaware, Susquehanna, and NYC water supply systems, and is currently engaged in modeling the Appalachicola-Chattahoochee-Flint and Alabama-Coosa-Tallapoosa Basins for the Atlanta Regional Commission and the South Saskatchewan River Basin in Alberta.

Dr. Sheer has done pioneering work in the development of water resources modeling technology and the use of Computer Aided Negotiation and Operations Exercises. His work on HydroLogics' OASIS modeling system led to a U.S. Patent. He has received Best Journal Paper citations from both AWWA and ASCE, was a founding member of the National Research Council's Water Science and Technology Board, and serves on the NRC's Committee to review the Florida Keys Carrying Capacity Study.

Mark Simpson, Manatee County Manatee County Utilities Department

Mark Simpson is the Water Division Manager of the Utilities Department for the Manatee County Government, Manatee County, Florida. Mr. Simpson has Bachelor's degrees in Chemistry and Biology from the University of South Florida and has been with Manatee County for 26 years. He has worked in the Manatee County Utilities Department Quality Control Laboratory for the majority of his career, with major focus on researching the prevention and removal of algal by-products from potable source surface water. He is the author or co-author of over 25 technical papers, research reports, and presentations to professional conferences covering subjects including water quality, treatment, and laboratory techniques. He is member of the American Water Works Association and the North American Lakes Management Society. He is longtime member of the AWWA Taste and Odor Committee and of Standard Methods. He is also the proud father of four enchanting and brilliant daughters from the age of 10 to 22.

Joel Smith, Vice President, Stratus Consulting, Inc.

Joel Smith, Vice President with Stratus Consulting, has been analyzing climate change impacts and adaptation issues for over 20 years. Mr. Smith was a coordinating lead author for the synthesis chapter on climate change impacts for the Third Assessment Report of the Intergovernmental Panel on Climate Change and was a lead author for the IPCC's Fourth Assessment Report. He was recently nominated to be on the National Academy of Sciences "Panel on Adapting to the Impacts of Climate Change." He has provided technical advice, guidance, and training on assessing climate change impacts and adaptation to people around the world and for clients such as the EPA, the U.S. Agency for International Development, the U.S. Country Studies Program, the World Bank, the UN, a number of states and municipalities in the U.S., the Pew Center on Global Climate Change, the Electric Power Research Institute, the National Commission on Energy Policy, and the Rockefeller Foundation. He worked for the EPA from 1984 to 1992, where he was the deputy director of Climate Change Division. He is a coeditor of EPA's Report to Congress: *The Potential Effects of Global Climate Change on the United States*, published in 1989; *As Climate Changes: International Impacts and Implications*, published by Cambridge University Press in 1995; and *Adaptation to Climate*

Proceedings of the First National Expert and Stakeholder Workshop on
Water Infrastructure Sustainability and Adaptation to Climate Change

127

Change: Assessments and Issues, published by Springer-Verlag in 1996, Climate Change, Adaptive Capacity, and Development, published in 2003 by Imperial College Press, and The Impact of Climate Change on Regional Systems: A Comprehensive Analysis of California published in 2006 by Edward Elgar. He joined Hagler Bailly in 1992 and Stratus Consulting in 1998. He has published more than thirty articles and chapters on climate change impacts and adaptation in peer-reviewed journals and books. Besides working on climate change issues at EPA, he also was a special assistant to the assistant administrator for the Office of Policy, Planning and Evaluation. He was a presidential management intern in the Office of the Secretary of Defense from 1982 to 1984. He has also worked in the U.S. Department of Energy and the U.S. Agency for International Development. He received a BA (*magna cum laude*) from Williams College in 1979, and a Master's in Public Policy from the University of Michigan in 1982.

Dr. Neil Stiber, EPA Office of the Science Advisor
Neil A. Stiber is an environmental scientist in the EPA's Office of the Science Advisor (OSA) where he is Interim Special Assistant for the Chief Scientist. Upon joining the EPA in 2003, Dr. Stiber worked in the Office of Research and Development (ORD) with the Council for Regulatory Environmental Modeling (CREM) where he was a co-author of the Guidance on Environmental Models and the primary developer of the CREM Models Knowledge Base. Next, he served on the Program Support Staff of ORD's Office of Science Policy where he focused on waste, contaminated sites, asbestos, and brownfields. Following that, he worked as staff to the Science Policy Council (SPC). While at the SPC, he promoted collaboration among agency-wide and inter-agency asbestos workgroups, supported the Expert Elicitation Task Force, coordinated activities between EPA and the NAS, and worked on many issues at the nexus of science policy, including climate change. Prior to joining the EPA, he worked for several years as a consultant specializing in environmental risk assessment, site investigation, and remediation. He received a B.S. in civil engineering from Duke University, a M.S. in civil engineering from Northwestern University, and M.S & Ph.D. from the Department of Engineering and Public Policy at Carnegie Mellon University. His research interests include expert elicitation and environmental decision making.

James Taft, Executive Director, Association of State Drinking Water Administrators
James Taft is the Executive Director of the Association of State Drinking Water Administrators (ASDWA). The Association supports state drinking water programs in their various efforts to ensure safe drinking water for the American public. Mr. Taft has over 30 years experience in water and wastewater policy and technical issues. Prior to joining the Association in 2003, he worked for the EPA (in the Office of Ground Water and Drinking Water and the Office of Wastewater Management), the U.S. Agency for International Development (in Central and Eastern Europe), the Virginia Department of Environmental Quality, the Ocean County (New Jersey) Utilities Authority, and the Ohio River Valley Water Sanitation Commission. He has a B.S. in Biology from Villanova University and an M.S. in Environmental Engineering from the University of Cincinnati.

Claudio Ternieden, Water Environment Research Foundation
Claudio Ternieden helps direct the research efforts of the Water Environment Research Foundation, a nonprofit organization focused on the science and technology of water and wastewater management. Mr. Ternieden helps lead WERF's climate change, energy, wastewater treatment operations and optimization research efforts and works with federal, state and local agencies, academia and the private sector to seek solutions to municipal challenges affecting water quality. Previously, he worked on climate change issues in the aviation and transportation industry and contributed to the National Academy of Sciences Transportation Research Board's environmental efforts. He also worked in environmental regulatory issues at the US Environmental Protection Agency and the Indiana Department of Environmental Management. He has a law degree and a certificate in Environmental Law from Pace University School of Law, in White Plains, NY; a BA from

Proceedings of the First National Expert and Stakeholder Workshop on
Water Infrastructure Sustainability and Adaptation to Climate Change

128

Concordia College, Bronxville, NY; and graduate work in Public Policy (Climate Policy) from George Mason University, Arlington, VA.

Brad Udall, University of Colorado – NOAA Western Water Assessment

Brad Udall is director of Western Water Assessment, one of seven RISA (Regional Integrated Sciences and Assessments) programs funded by the Office of Global Programs at NOAA. These programs are designed to develop partnerships with regional stakeholders and tailor NOAA data products to meet their needs. Lessons learned here are also contributing to NOAA's emerging "National Climate Service," the climate analog to the existing National Weather Service.

Alan Vicory, P.E., Ohio River Valley Water Sanitation Commission

Alan Vicory serves as Executive Director and Chief Engineer for the Ohio River Valley Water Sanitation Commission (ORSANCO). Appointed to the position in May 1987 after previous responsibilities with the Commission staff as Environmental Engineer and Manager of Technical Services, Mr. Vicory directs the programs of the Commission, which include establishment of regulatory requirements for discharges, water quality and biological monitoring systems, detection and response to spills, applied research, coordination of states and federal programs and public education and involvement. ORSANCO, known worldwide for its accomplishments in water pollution control on a watershed basis, was established in 1948 by state compact. Members of the Commission represent Illinois, Indiana, Kentucky, New York, Ohio, Pennsylvania, Virginia, West Virginia and the United States. He received a B.S. degree in Civil Engineering from Virginia Military Institute in 1974. He is a Registered Professional Engineer and Board Certified in environmental engineering (water and wastewater) by the American Academy of Environmental Engineers. He is current Vice-chairman of the Board of the Water Environment Research Foundation (WERF) a former Chairman of the International Water Association's (IWA) Watershed and River Basin Management Specialist Group, and is currently a member of the Association's Strategic Council. He also is a Past President of the American Academy of Environmental Engineers (AAEE) and the Association of State and Interstate Water Pollution Control Administrators (ASIWPCA). He has provided contributions to published texts, published and presented numerous professional papers, has provided keynote remarks at several technical and professional conferences nationally and internationally, and has served on many expert panels.

Marc Waage, P.E., Denver Water / Water Utility Climate Alliance

Marc Waage manages Denver Water's integrated resource planning and climate change planning. Denver Water, the largest water utility in Colorado, serves 1.2 million people. Prior to managing planning, he managed Denver's water collection system operations. Before his 22-year career at Denver Water, he worked briefly for the Bureau of Reclamation and the Bureau of Indian Affairs on agricultural irrigation projects. He has a Bachelor's degree (with high distinction) and a Master's degree in Civil Engineering from Colorado State University and is a professional engineer. One of his favorite activities is recreating in Denver's mountain watersheds.

Laura Wharton, King County Department of Natural Resources and Parks

Laura Wharton is the Supervisor of Comprehensive Planning and Asset Management Development for the King County Wastewater Treatment Division. Ms. Wharton and her staff are responsible for near term and long-range planning for the wastewater conveyance system and treatment plant infrastructure for the greater Seattle Metropolitan area, Combined Sewer Overflow control planning within the City of Seattle and reclaimed water planning for the regional treatment system. She has twenty-five years of experience leading major capital planning and siting projects for government agencies. She has a Bachelor's of Science degree in Forestry from Michigan Technological University.

Proceedings of the First National Expert and Stakeholder Workshop on
Water Infrastructure Sustainability and Adaptation to Climate Change

129

Dr. Pai-Yei Whung, Chief Scientist, EPA Office of the Science Advisor
As Chief Scientist, Dr. Pai-Yei Whung shares fully with the EPA Science Advisor in planning, developing, and implementing cross-Agency scientific efforts. This includes providing program management and technical support to the Science Advisor by independent scientific opinions and through leading OSA staff and its multiple science-policy functions. Dr. Whung has a doctoral degree in climate change, marine and atmospheric chemistry, a master's degree in oceanography and marine chemistry, and a bachelor's degree in oceanography and geology. She has fifteen years of field research experience and eight years of program management and leadership in bioenergy, air quality, water quality, weather, sustainable ecosystems, climate change, and agricultural research. Her research has been published in peer-reviewed journals and presented at many professional meetings. Prior to joining EPA, she worked in the Agricultural Research Service at UDSA and for NOAA where she had a detail to the World Meteorological Organization. Through these positions she has cultivated a broad perspective on science in the federal government. In these positions, her experiences with EPA included conducting water quality research in bays, developing analytical techniques, and initiating interagency science programs. She has successfully worked with the Office of Management and Budget (OMB), the Office of Science and Technology (OSTP), Congress, and private-sector stakeholders on scientific initiatives. In addition, she has led the development of several policy documents with multiple federal and state agencies, governors associations and universities (notably, the development of the National Science and Technology Council Subcommittee on Disaster Reduction's strategic action plan for implementation of a National Integrated Drought Information System.)

Dr. Y. Jeffrey Yang, P.E., EPA National Risk Management Research Laboratory
Jeff Yang is an environmental scientist with the EPA National Risk Management Research Laboratory stationed in Cincinnati, Ohio. Dr. Yang has a broad range of professional knowledge and research experience in water resources, drinking water, wastewater, groundwater and storm water engineering and management. In his 26 years of professional career, he spent a half of the time in private practice on large engineering projects and program management before coming back to the research side. He has a bachelor's degree, two master's degrees, and a Ph.D. degree from China and U.S. He is a licensed professional engineer and professional geologist in the states and a Diplomat of Water Resources Engineering (D.WRE) in the ASCE's AAWRE. At the EPA, he has enjoyed in developing the Water Resources Adaptation Program (WRAP), participating in Agency's research and rule-making activities, having served as an ad-hoc peer reviewer for several international journals and in several professional committees. He also has published extensively.

Dr. David Yates, National Center for Atmospheric Research
David Yates is a Project Scientist in the Research Applications Laboratory at the National Center for Atmospheric Research, Boulder Colorado and Research Associate with the Stockholm Environment Institute's US Center in Davis, CA. Dr. Yates research has focused both on local scale hydrologic problems (flash floods, land use-land cover, climate change), as well as climate change impacts on water and agricultural systems. He is PI on an EPA Office of Research and Development Project which is developing an analytic tool- the Water Evaluation and Planning model- for looking at the combined effects of climate change and land-use on ecological resources and freshwater services. This tool was partially developed with funding from the AWWA Research Foundation, to help water utilities with long-range planning that includes climate change impacts. With Kathleen Miller and support from the AWWA Research Foundation, he has helped develop an educational primer for use by the drinking water utility industry that outlines the current state of scientific knowledge regarding the potential impacts of global climate change on water utilities, including impacts on water supply, demand and relevant water quality characteristics.

Proceedings of the First National Expert and Stakeholder Workshop on
Water Infrastructure Sustainability and Adaptation to Climate Change

130

www.ingramcontent.com/pod-product-compliance
Lightning Source LLC
Chambersburg PA
CBHW080641180526

45168CB00008B/3252